36

Topics in Heterocyclic Chemistry

For further volumes:
http://www.springer.com/series/7081

Aims and Scope

The series Topics in Heterocyclic Chemistry presents critical reviews on present and future trends in the research of heterocyclic compounds. Overall the scope is to cover topics dealing with all areas within heterocyclic chemistry, both experimental and theoretical, of interest to the general heterocyclic chemistry community.

The series consists of topic related volumes edited by renowned editors with contributions of experts in the field.

Janine Cossy
Editor

Synthesis of Saturated Oxygenated Heterocycles II

7- to 16-Membered Rings

With contributions by
Martin Cordes • Markus Kalesse • José Marco-Contelles •
Olivier Piva • Elena Soriano

Springer

Editor
Janine Cossy
ESPCI
Laboratory of Organic Chemistry
Paris Cedex 05
France

ISSN 1861-9282 ISSN 1861-9290 (electronic)
ISBN 978-3-642-41469-5 ISBN 978-3-642-41470-1 (eBook)
DOI 10.1007/978-3-642-41470-1
Springer Heidelberg New York Dordrecht London

Library of Congress Control Number: 2014931980

Printed on acid-free paper

Springer is part of Springer Science+Business Media (www.springer.com)

Preface

Two volumes are dedicated to the synthesis of saturated oxygenated heterocycles and consist of eight chapters covering the synthesis of 5- to 16-membered ring cyclic ethers and lactones. Rather than offer an exhaustive description of the synthesis of cyclic ethers and lactones, these volumes present methods and strategies to synthesize heterocycles and thus helping the reader to find suitable methods for obtaining a desired saturated oxygenated heterocycle. The first volume comprises five chapters and the second volume three chapters.

In chapter entitled "Synthesis of Substituted Tetrahydrofurans," J. D. Rainier outlines the advances that have been made during the last 10 years in the synthesis of tetrahydrofurans such as nucleophilic additions to acetals and hemiacetals, cycloadditions, oxidative cyclizations, furan reductions, Prins-pinacol cascades, ring-opening of bicyclic substrates, and nucleophilic substitutions.

In chapter "Synthesis of Saturated Tetrahydropyrans," S.D. Rychnovsky, M. A. Perry, and N. Sizemore review the common strategies to access tetrahydropyrans such as the formation of O1–C2, C2–C3, C3–C4, O1–C6, and C2–C3 bonds, as well as C2 functionalization of lactols and lactones.

The chapter "Synthesis of Saturated Six-Membered Ring Lactones" by K. P. Kaliappan and K. Palanichamy describes various selected methods such as lactonization of δ-hydroxy acid derivatives, oxidation, electrophilic cyclization, intramolecular nucleophilic displacement, radical and reductive cyclizations, palladium-catalyzed lactonization, as well as carbonylation and carboxylation.

The synthesis of 7-oxabicyclo[2.2.1]heptanes and derivatives is reported in the chapter "Synthesis of 7-Oxabicyclo[2.2.1]heptane and Derivatives" has been written by P. Vogel and A. J. Moreno-Vargas. Most of the methods, reported to access 7-oxabicyclo[2.2.1]heptanes, are Diels–Alder reactions, but non-Diels–Alder reactions such as electrophilic cyclizations have also been included in this chapter. As 7-oxabicyclo[2.2.1]heptane derivatives can be good precursors of other oxygenated heterocycles, their ring cleavage either by cleavage of a C-O or a C-C bond have been reported. In addition, as 7-oxabicyclo[2.2.1]heptane derivatives are extremely useful synthons, few syntheses of natural products and bioactive compounds, using these synthons, have been described.

In chapter "Synthesis of 5,6- and 6,6-Spirocyclic Compounds," M. A. Brimble and L. A. Stubbing describe a number of recently reported and useful methods to synthesize 5,6- and 6,6-spirocyclic compounds, including their applications to the synthesis of natural products and bioactive compounds containing spiroacetal scaffolds. One can find dehydrative spirocyclization of dihydroxyketones, metal-catalyzed addition/elimination of allylic alcohols, acid-catalyzed spirocyclization of hemiacetals, spirocyclization of exo- and endocyclic enol ethers, transition-metal-catalyzed hydroalkoxylation of alkynes, electrophilic cyclization and oxa-Michael cyclization, intramolecular hetero-Michael addition, ring-opening of epoxides and cyclopropanes, cycloadditions, furan oxidation, intramolecular hydrogen abstraction, reductive cyclizations, ring-closing metathesis, and rearrangements.

In chapter "Synthesis of Seven-Membered-Ring Ethers and Lactones," O. Piva describes the access to saturated oxygenated 7-membered cyclic ethers, by ring expansion of oxygenated structures, by formation of C–O and C–C bonds using different methods. For 7-membered cyclic lactones, oxidative processes, halolactonization, lactonization of ω-hydroxyacids, tandem Suzuki coupling and lactonization, and ring enlargement are reported.

For the synthesis of 8- to 10-membered ring ethers, in chapter "Synthesis of Eight- to Ten-Membered-Ring Ethers," J. M. Contelles and E. Soriano focus on the formation of carbon–carbon double bonds by metathesis, as well as on the formation of carbon–carbon single bonds. The authors also report on the cyclization to form C–O bonds, ring expansion, ring-opening, and rearrangement.

Chapter "Synthesis of 12- to 16-Membered-Ring Lactones" is dedicated to the synthesis of 12- to 16-membered ring lactones. In this chapter, M. Kalesse and M. Cordes present an overview of the macrocyclization of seco-acids as well as new effective procedures to access 12- to 16-membered ring lactones such as ring-closing metatheses of alkynes and olefins. The authors also report the use of ketene sources and benzodioxinones to produce macrocyclic lactones. Nitrile oxide-olefin cycloaddition, intramolecular C–H oxidative macrolactonization, and Yamaguchi and Mukaiyama macrocyclization as well as macrolactonization via thioester or using phosphorus reagents are described.

I would like to express my sincere gratitude to all the authors of these chapters for their efforts and outstanding contributions. I would also like to thank B. Maes for giving me the opportunity to edit these two volumes, to Elizabeth Hawkins and Tanja Jaeger from Springer for coordinating the project, and to Fairin Miriam John Bennet for the editing process, for her help and patience.

Finally, I hope that this book will be a good source of inspiration for those planning the synthesis of saturated oxygenated heterocycles, for solving specific synthetic problems, or for elaborating on new synthetic tools.

Paris, France Janine Cossy

Contents for Volume I

Contents for Volume II

Synthesis of Seven-Membered Ring Ethers and Lactones

Olivier Piva

Contents

O. Piva (✉)
Institut de Chimie et de Biochimie Moléculaires et Supramoléculaires, Université Lyon 1, UMR CNRS 5246, 43, Bd du 11 novembre 1918, 69622 Villeurbanne, France
e-mail: piva@univ-lyon1.fr

J. Cossy (ed.), *Synthesis of Saturated Oxygenated Heterocycles II*, Topics in Heterocyclic Chemistry 36, DOI 10.1007/978-3-642-41470-1_1,

Abstract Compared to parent oxygenated five- and six-membered rings, oxepanes represent a rare structure found in Nature. Nevertheless, natural products with significant biological activities possess either only one seven-membered ring or one or multiple oxepane subunits fused to other oxygenated heterocycles. Therefore, a large number of methods have been designed to access these structures with a special attention on metal-catalysed coupling reactions.

Keywords Cyclisation • Epoxides • Lactones • Metal-induced cyclisations • Metathesis • Oxepanes • Oxepins • Oxidation • Palladium-catalyzed coupling • Prins reaction • Radical cyclisation • Ring enlargement • Ring opening • Samarium diiodide

1 Introduction

Compared to the corresponding five- and six-membered cyclic ethers, the homologous oxepanes and oxepins are less common in Nature. Some of them depicted in Figs. 1 and 2 exhibit a large variety of pharmaceutical properties. Therefore, due to their usually low abundance, synthetic chemists have developed a large panel of methods to reach them. In this chapter, the access to various seven-membered oxacycles including oxepanes, oxepins or benzoxepines will be described, and, to a less extent, the synthesis of polycyclic structures reported in different reviews [1–6].

2 Ring Expansion of Oxygenated Structures

For steric and also entropic reasons, the direct formation of seven-membered rings from acyclic structures is usually difficult. Therefore, the ring expansion of dihydropyranes into oxepanes has emerged as a powerful and high-yielding method.

2.1 Cyclopropanation and Ring Opening of Bicyclo[4.1.0]heptanes

The synthesis of heptanoside **2** has been achieved according to a two-step procedure. D-Glucal derivatives can be easily converted to bicyclo[4.1.0]heptanes by a diastereoselective addition of a dibromocarbene to the double bond. By heating **1** in toluene, in the presence of silver acetate, the cleavage of a carbon–bromide bond occurred and is followed by an electrocyclic ring opening. The corresponding carbocation was next trapped by various nucleophiles. The best results were obtained with sodium acetate. Noteworthy, in refluxing toluene and in the absence

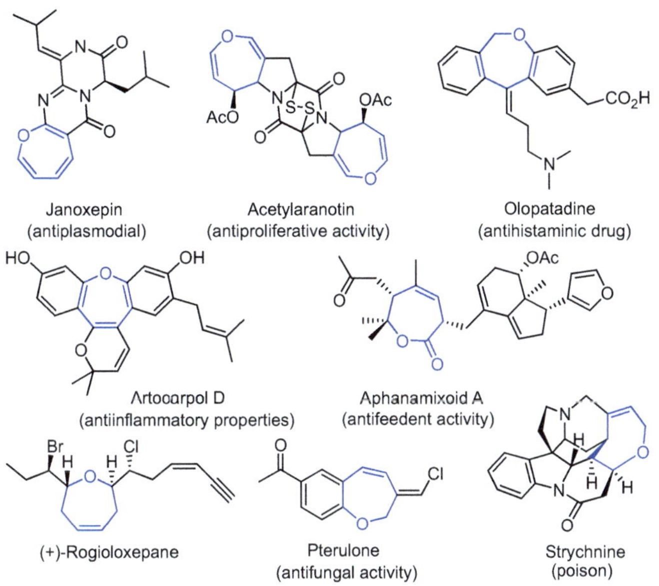

Fig. 1 Oxepins, benzoxepines, oxepenes and related structures

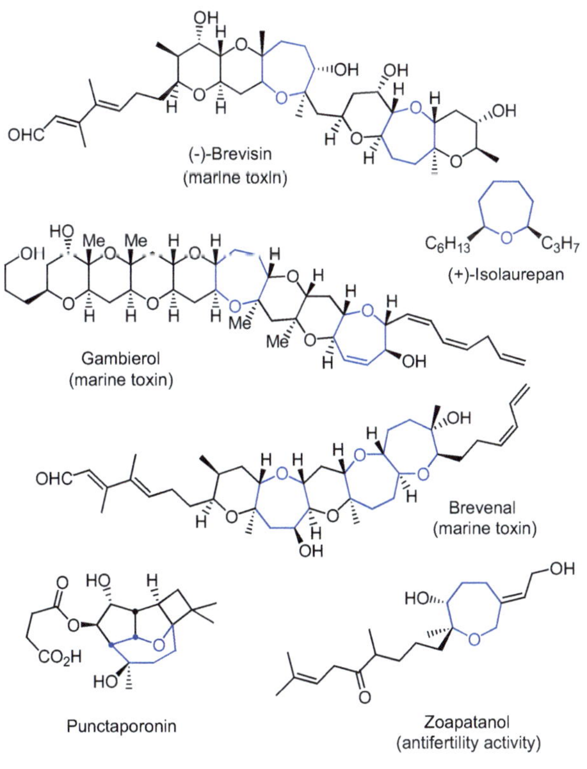

Fig. 2 Natural-occurring oxepanes and polycyclic ethers

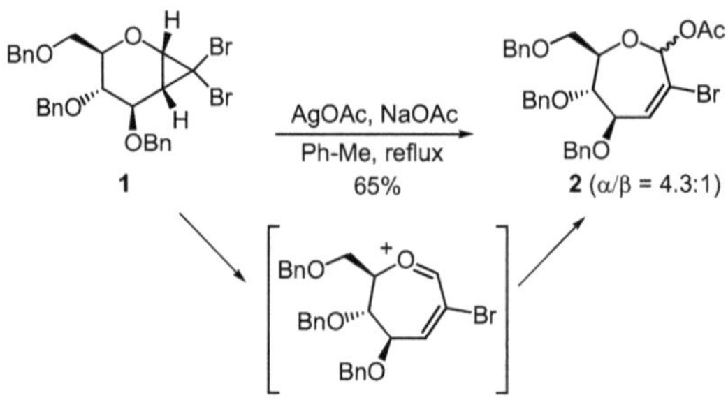

Scheme 1 Ring opening of dibromo-bicyclo[4.1.0]heptanes

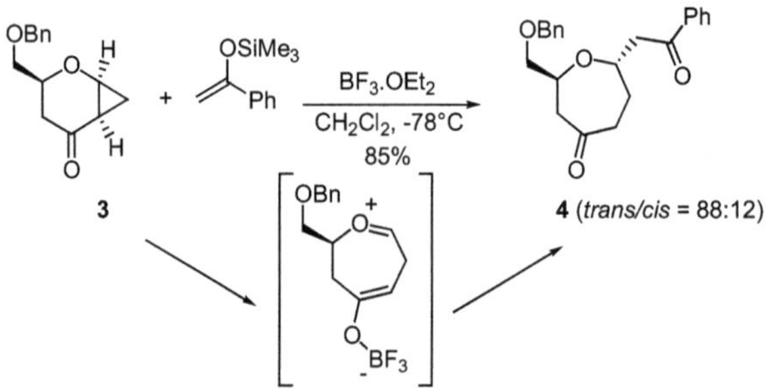

Scheme 2 Ring opening of bicyclo[4.1.0]heptanone

of silver salts, **1** was transformed to a 2-*C*-methylene α-pyranoside [7, 8] (Scheme 1).

A ring expansion of cyclopropylpyranone **3** was observed when treated in the presence of $BF_3{\cdot}OEt_2$ at −78 °C [9, 10]. Enol ethers have been used as nucleophiles to quench the oxonium intermediate to deliver 2,6-disubstituted 4-oxo-oxepane **4** with a high *trans* selectivity, resulting to the minimisation of the steric hindrance during the nucleophilic attack (Scheme 2).

2.2 Ring Expansion of Pyranones by Arndt–Eistert Reaction

In the course of the total synthesis of gambierol by Mori et al. [11, 12], the formation of the E ring was achieved by the ring expansion of β-pyrone **5** in 52 % yield, when submitted to trimethylsilyldiazomethane and $BF_3{\cdot}OEt_2$ (Scheme 3).

This process was also considered to access to a second oxepane subunit (H ring) present in the same target but with a far better yield (81 %).

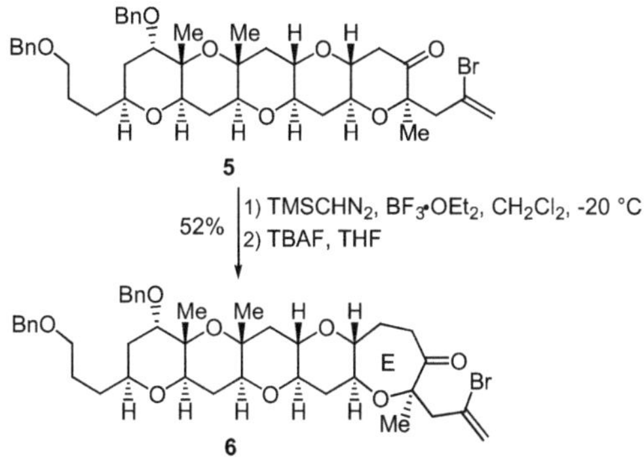

Scheme 3 Arndt–Eistert reaction applied to the access of E ring of gambierol

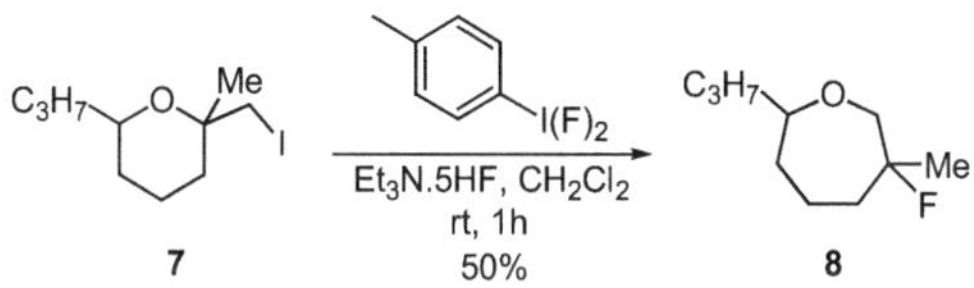

Scheme 4 Ring expansion of 2-halomethyl tetrahydropyran to 3-fluorooxepane

2.3 Ring Expansion of 2-Halomethyl Cyclic Ethers

The ring expansion of readily available iodomethyl tetrahydropyrans such as **7**, promoted by hypervalent iodine reagents, has been extensively studied by Hara's group (Scheme 4). Using *p*-iodotoluene difluoride, a rare 3-fluorooxepane derivative **8** was obtained, however, in a moderate yield [13, 14].

2.4 Other Ring Expansions

A two-step synthesis of 2,3-dihydrobenzooxepine **14** was recently realised from salicylaldehyde and ethyl chloroacetate (Scheme 5) [15]. At first, a Knoevenagel/hemiketalisation afforded 2-(chloromethyl)-2*H*-chromen-2-ol **10** which was not isolated but directly treated with a tertiary amine such as the Hünig base. Under these basic conditions, the ketal underwent a ring opening to deliver intermediately the corresponding α-chloroketone **12** which reacted with the generated phenolate according to an S_N2 process. The resulting benzoxepinone **13** was finally involved in a classic Wittig reaction to afford **14** with a yield of 96 %.

The oxidative ring expansion of 4-methylene-chromanes was also realised (Scheme 6). For example **17**, prepared from **15** according to a two-step Mitsunobu reaction/intramolecular Heck reaction sequence, was submitted to Koser and Justik reagent [PhI(OH)OTs] to produce benzo[*b*]oxepin **18** in good yield (64 %) [16].

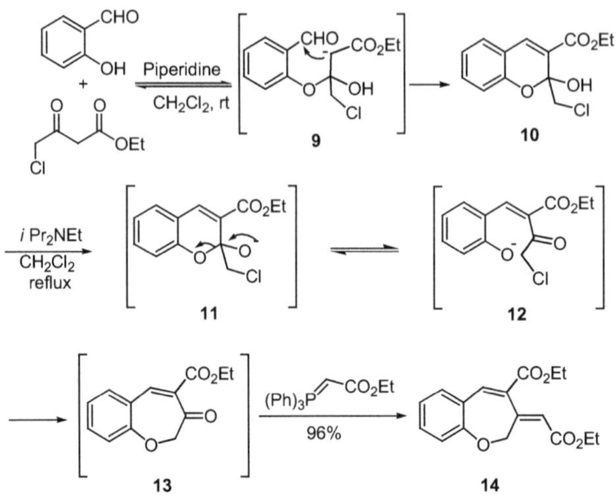

Scheme 5 Access to benzoxepinone by condensation of salicylaldehyde and ethyl chloroacetate

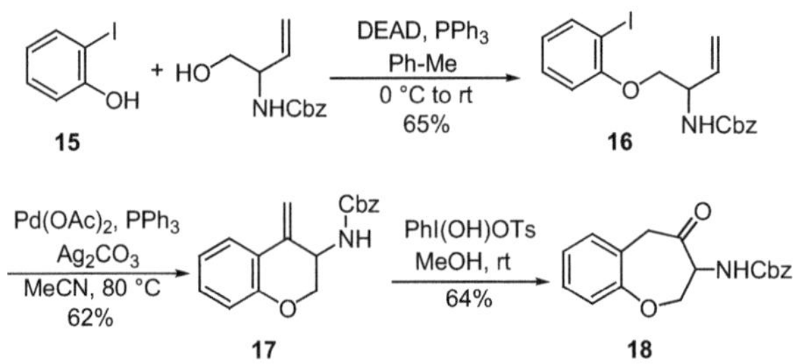

Scheme 6 Ring expansion of 4-methylene-chromanes

3 Formation of C–O Bonds

3.1 Cyclisations Involving Oxygenated Nucleophiles

3.1.1 Formation of Oxepanes Under Acidic Conditions

1,6-Diols can be transformed into oxepanes by treatment with various proton sources. For example, sorbitol in refluxing toluene, in the presence of a catalytic amount of TfOH, furnished the septanoside **20** in 79 % without epimerisation of the stereogenic centres (Scheme 6). Surprisingly, no formation of five- or six-membered rings was noticed. Similar results were obtained with other proton sources such as TsOH (Scheme 7) [17].

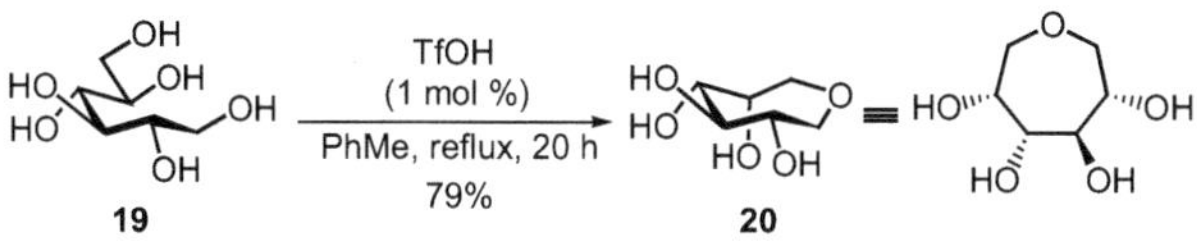

Scheme 7 Cyclisation of sorbitol into oxepane **20** under acidic conditions

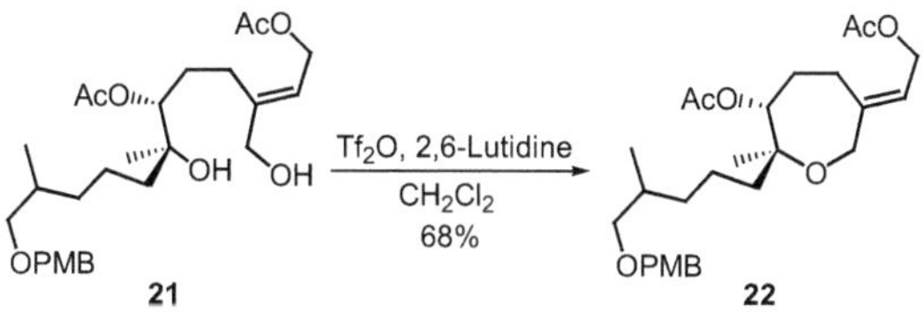

Scheme 8 Direct access to the core structure of (+)-zoapatanol

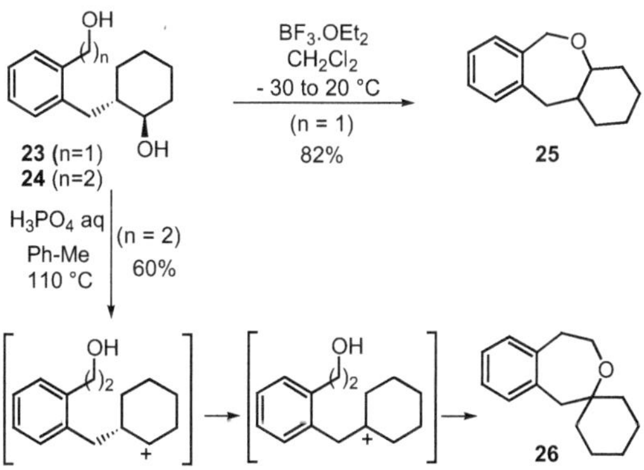

Scheme 9 Distinct cyclisation pathways of benzylic and homobenzylic alcohols into benzoxepines

A related process applied to **21** has been used as the key step in the total synthesis of (+)-zoapatanol, reported by Raghavan and Babu. Four additional steps were required to reach the final target (Scheme 8) [18].

Tetrahydrobenzoxepine **25** was similarly obtained from benzylic alcohol **23** (n = 1), when treated with $BF_3{\cdot}OEt_2$ [19]. Interestingly, homobenzylic structure **24** (n = 2) did not lead to the eight-membered ring ether but to the spiranic oxepane **26**. These results were rationalised by considering a Wagner–Meerwein rearrangement of a secondary to a tertiary carbocation (Scheme 9).

Under acidic conditions, biphenyl tertiary alcohol **27** gave access to dibenzoxepin **28** in very good yield [20] (Scheme 10).

Cyclisation of unsaturated alcohols has been also evaluated to synthesise oxepanes. Treatment of **29** with the amberlyst-15 resin allowed the exclusive formation of the seven-membered ring ether **30** [21]. The regioselectivity observed was unambiguously governed by the formation of the tertiary carbocation (Scheme 11).

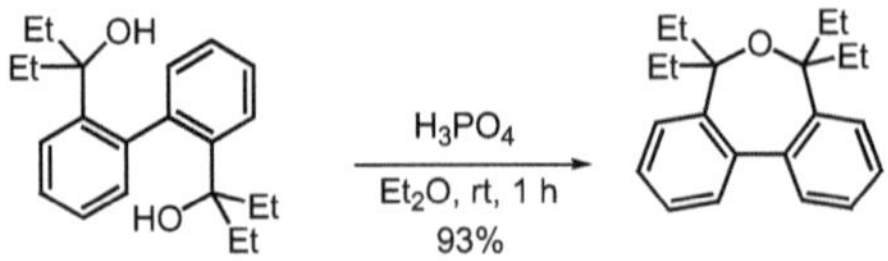

27 28

Scheme 10 Formation of a dibenzoxepin from a dibenzylic alcohol under acidic conditions

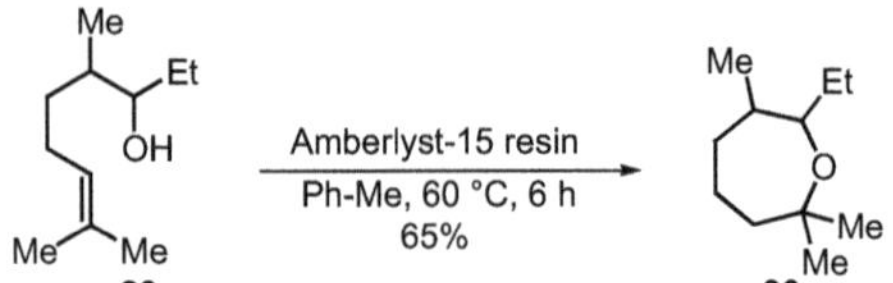

29 30

Scheme 11 Formation of an oxepane from an unsaturated alcohol

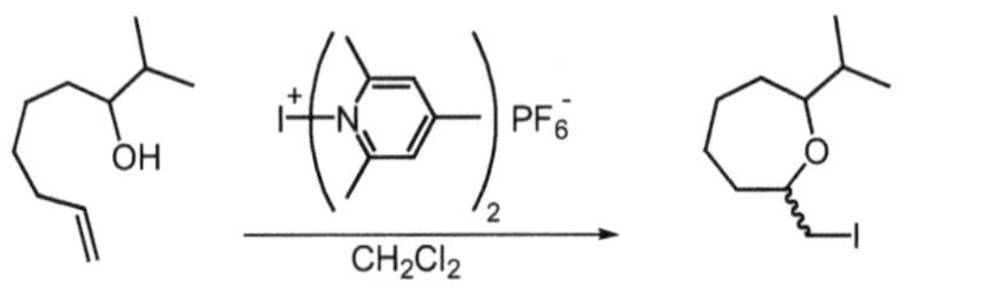

31 **32** (*cis/trans* = 50:50)

Scheme 12 Formation of a 2-iodomethyl-oxepane by iodoetherification

3.1.2 Haloetherification

Molecular iodine and bis(*sym*-collidine)iodine are the electrophiles of choice to furnish oxepanes from hept-6-en-1-ols (Scheme 12).

Unfortunately, in the absence of additional stereogenic centres already present on the chain, no stereocontrol was observed [22].

3.1.3 Metal-Catalyzed Cycloisomerisation

In the course of the synthesis of *epi*-dithiodiketopiperazines, a class of metabolites recently isolated from fungi, Reisman et al. reported a two-step sequence to the dihydrooxepine subunit. They found that rhodium salts catalysed the 7-*endo* heterocyclisation of the highly functionalised acetylenic hydromethylpyrrolidine **33** into the bicyclic structure **34** which underwent β-elimination to the core structure of (−)-acetylaranotin (Scheme 13) [23].

A similar cyclisation has been also reported with acetylenic alcohols readily prepared from protected sugars and by using a stable tungsten Fischer carbene (Scheme 14) [24, 25].

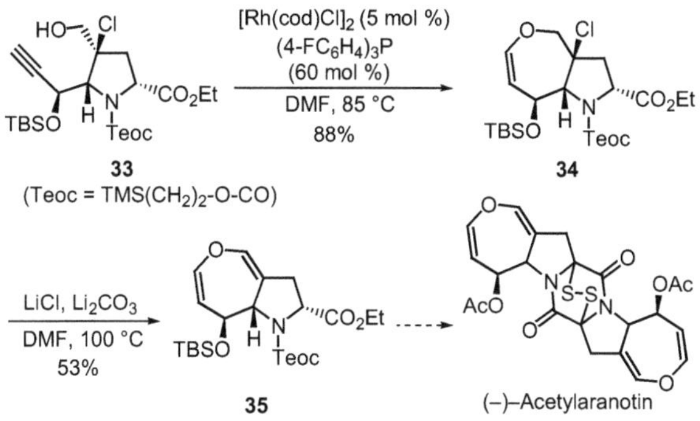

Scheme 13 Cyclisation of an acetylenic alcohol into a dihydrooxepine in presence of rhodium salts

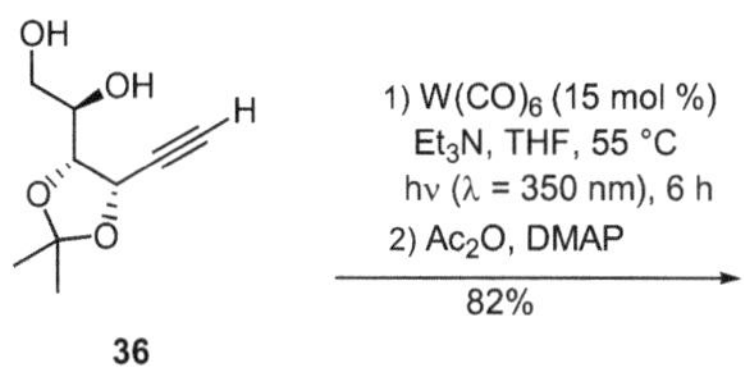

Scheme 14 Cycloisomerisation of acetylenic alcohol catalysed by tungsten hexacarbonyl

3.2 Ring Opening of Epoxides and Oxidation of Polyenes

Ring opening of epoxides is a common process to access oxepanes. For example, Suzuki et al. have shown that the combination of $(Bu_3Sn)_2O$ and a Lewis acid (L.A.) allowed the exclusive formation of the seven-membered ring **39** from epoxyalcohol **38** (Scheme 15) [26].

As reported previously, the formation of a transient stannyl ether enhanced the nucleophilicity of the oxygen atom, while the Lewis acid favoured the cleavage of the strained oxirane subunit. A direct application of this method has been used to synthesise the marine natural product (+)-rogioloxepane (Scheme 16) [27].

McDonald et al. has devised an alternative method, based on the ring opening of bis-oxiranes such as **42** through a cascade process to generate an oxepane ring (Scheme 17). Due to the Lewis acid assistance, the oxepane ring was formed according to a high regioselective *endo-tet* pathway [28, 29].

The opening of polyepoxide **44** was also realised to furnish compound **45** with up to four fused oxepane rings however in a low yield (12 %) (Scheme 18) [30].

Symmetrical disubstituted *trans*-dihydroxy oxepanes have been obtained from 1,7-dienes by treatment with RuO_4 generated from low-valence ruthenium species with sodium periodate (Scheme 19). Formation of a ruthenium(VI) diester was supposed at first, followed by a *trans* selective [3+2]-cycloaddition [31]. Only a catalytic amount of ruthenium salts was necessary, as $NaIO_4$ was able to regenerate the active RuO_4 species.

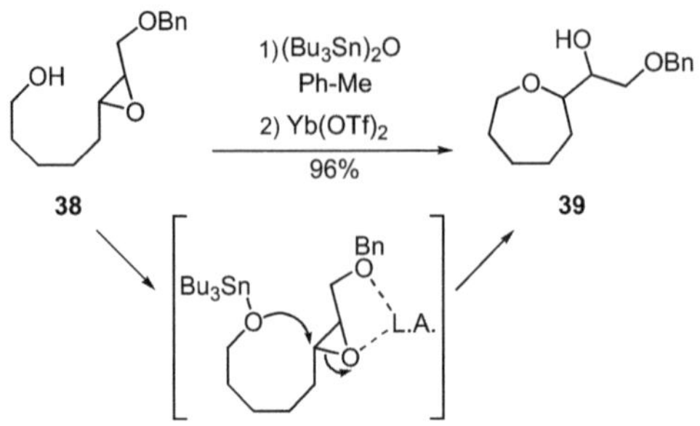

Scheme 15 Lewis acid catalysed intramolecular ring opening of an epoxyalcohol

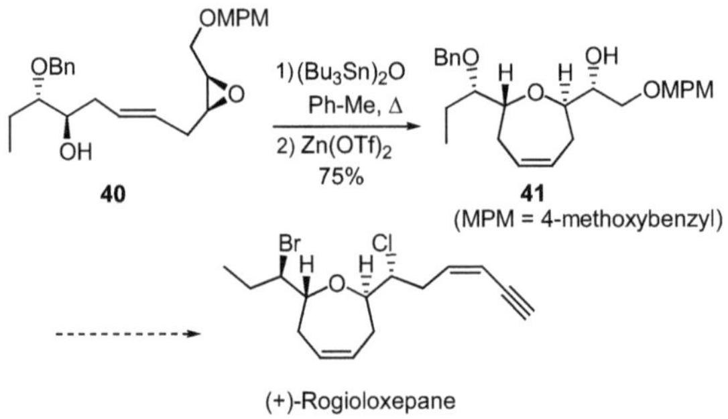

Scheme 16 Access to the core structure of rogioloxepane by intramolecular ring opening of epoxide **40**

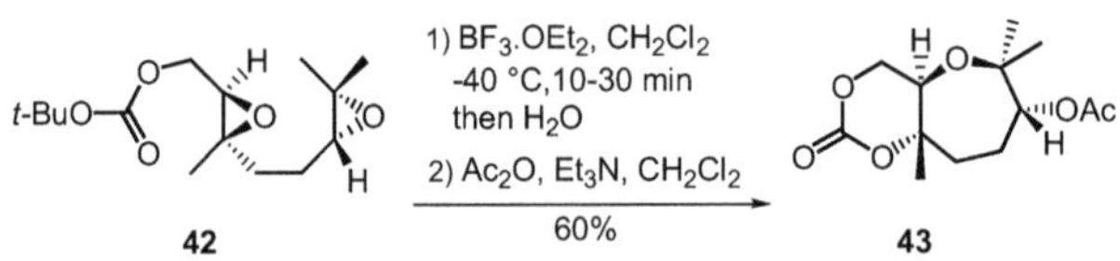

Scheme 17 Double-epoxide ring opening in a cascade process promoted by an L.A.

3.3 *Reductive Cyclisation of Hydroxyl Ketones*

Access to cyclic ethers can be easily achieved by reductive cyclisation of hydroxyketones. Combination of Et_3SiH with a strong Lewis acid has been widely used for this purpose [32]. Tripathi and Kumar have used these conditions to synthesise (+)-isolaurepan **49** (Scheme 20). Starting from the enantioenriched starting material **48**, they noticed the exclusive formation of the 2,7-disubstituted oxepane **49** in 84 % yield [33–35].

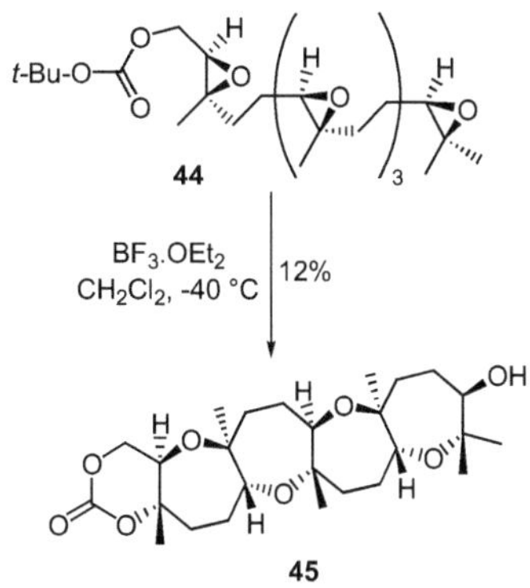

Scheme 18 Formation of four fused oxepanes by a cascade process

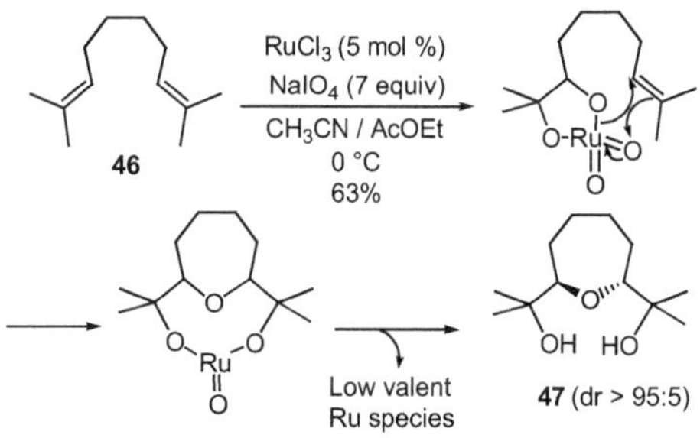

Scheme 19 Formation of 2,7-disubstituted oxepane by oxidative cyclisation of 1,7-dienes

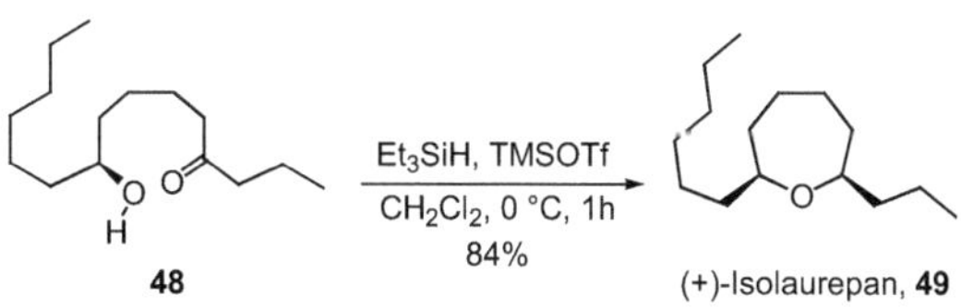

Scheme 20 Direct synthesis of (+)-isolaurepan by reductive cyclisation of a hydroxyl ketone

3.4 Ullmann Coupling

Coupling between a phenoxide and a bromoaryl derivative catalysed by copper(I) salts led efficiently to biarylethers. The intramolecular version of this process has been used for the synthesis of bulbophylol-B **51**, a dihydrodibenzo[*b*,*f*]oxepin with significant cytotoxic properties (Scheme 21) [36].

Under basic conditions, a tandem Knoevenagel reaction/Ullmann coupling has been devised to synthesise dibenzooxepin **53** from 2-hydroxybenzonitrile and 2-bromobenzaldehyde **52** in high yield (98 %) (Scheme 22) [37].

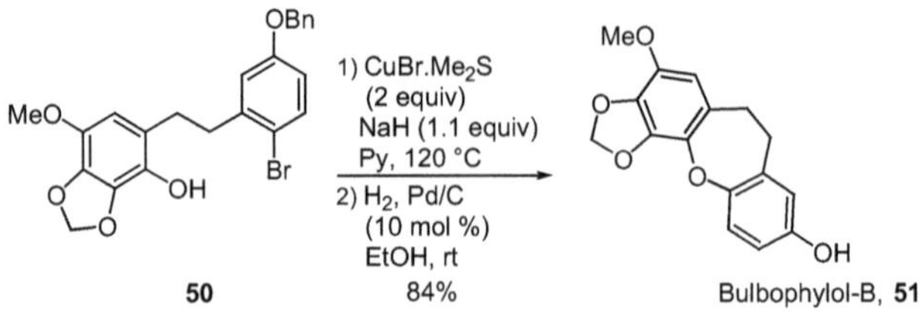

Scheme 21 Formation of a dihydrodibenzo[*b*,*f*]oxepin by an intramolecular Ullmann coupling

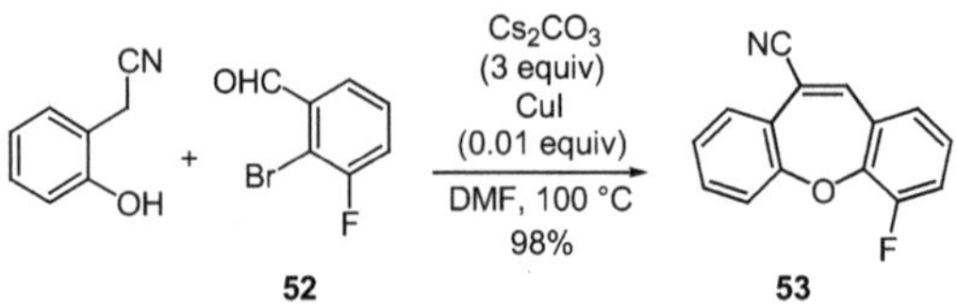

Scheme 22 Tandem Knoevenagel condensation/Ullmann coupling

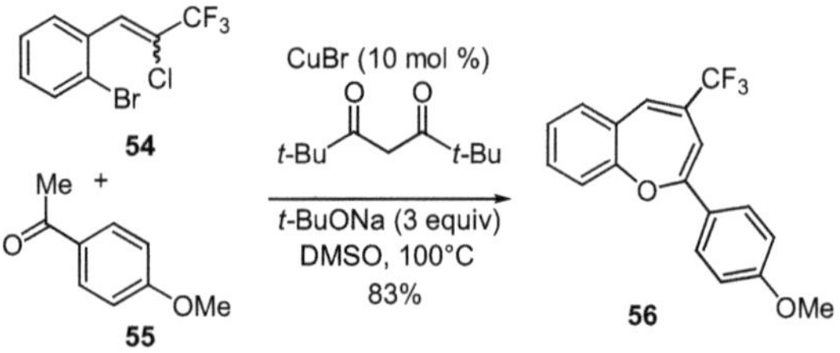

Scheme 23 Access to oxepin **56** by an S_NAr reaction

The same group also reported a metal-free cyclisation which occurred at higher temperatures (130 °C); in this case, the reaction probably proceeds via an S_NAr mechanism [38]. A copper-catalysed tandem reaction has been designed from *o*-halo β-chloro-β-trifluoromethylstyrenes and ketones (Scheme 23). By heating **54** and **55** at 100 °C in DMSO, under basic conditions, 4-trifluoromethylbenzoxepine **56** was obtained in 83 % yield [39].

3.5 *Oxa-Michael Reactions*

Intramolecular oxa-Michael reactions have been considered to build oxygenated heterocycles and found direct application in the total synthesis of natural products. In connection with the total synthesis of hemibrevetoxin B (HBTX-B), Fall et al. have reported the formation of the CD ring subunit by an intramolecular oxa-Michael reaction. The acceptor **58** was accessible by photooxidation of furan **57** in methanol, followed by cleavage of an *endo*-peroxide. The 7-*exo*-trig anionic cyclisation occurred after selective cleavage of a silyl ether under classical conditions with

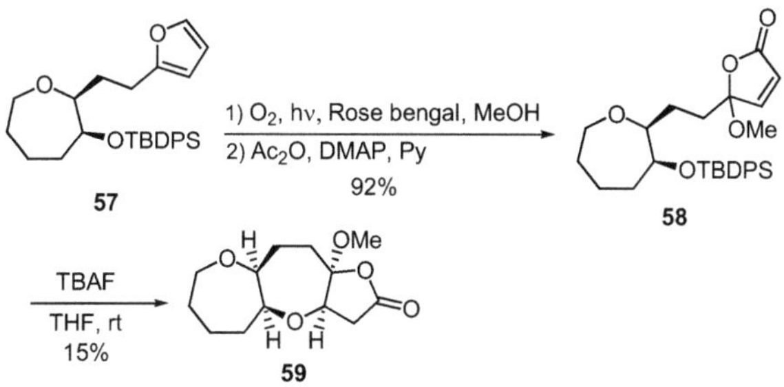

Scheme 24 Formation of a bis-oxepane by intramolecular oxa-Michael addition

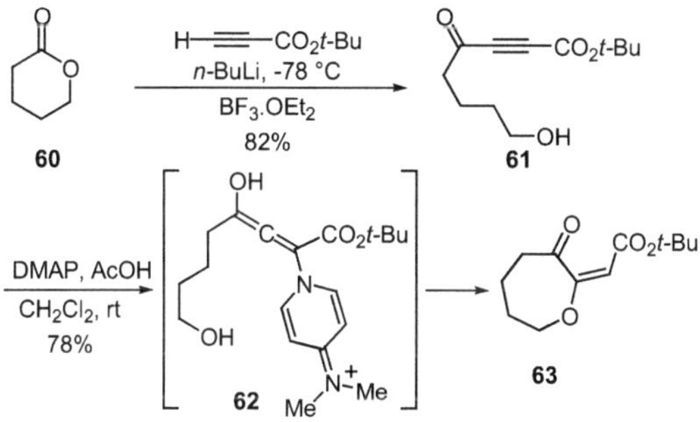

Scheme 25 Access to oxepanone by intramolecular oxa-Michael addition on a γ-keto-propiolate

fluorides (Scheme 24). Unfortunately, the yield was low (15 %), reflecting the difficulty for the molecule to adopt a suitable conformation for the cyclisation [40, 41].

An intramolecular oxa-Michael addition has been conducted on *tert*-butyl propiolate derivative **61**, directly obtained from α-pyrone **60**. The cyclisation occurred nicely in the presence of DMAP and acetic acid, to produce **63**, and an allene species **62** was suspected to be intermediately generated (Scheme 25) [42].

4 Formation of C–C Bonds

4.1 Electrophilic Cyclisations

4.1.1 Friedel–Crafts Reactions

Friedel–Crafts reaction was applied to 2-phenoxybenzoic acid **64** to access oxepanone **65**, precursor for the total synthesis of artocarpol D analogues (Scheme 26) [43].

Compound **67** was constructed from salicylaldehyde derivative **66** and *N*-methylindole, in the presence of various Lewis acids. It was noticed that only

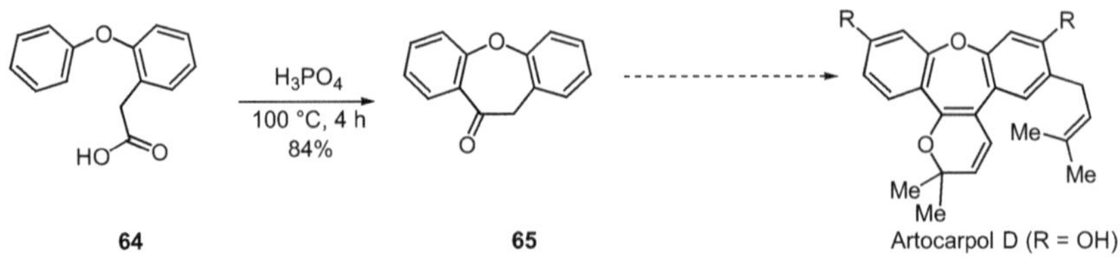

Scheme 26 Access to the core structure of artocarpol D by an intramolecular Friedel–Crafts reaction

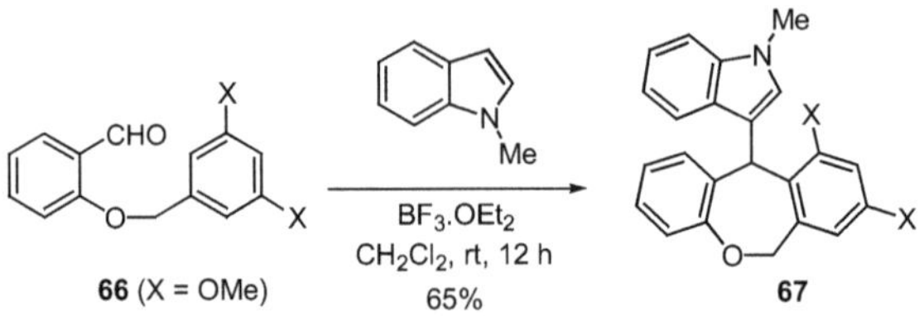

Scheme 27 Tandem indole condensation/Friedel–Crafts cyclisation

the electron-rich substrate (X = OMe) reacted under these mild conditions (Scheme 27) [44].

4.1.2 Prins Reaction and Related Processes

The Prins reaction is a very attractive process usually used to prepare six-membered heterocycles. The reaction performed from aldehydes and homoallyl alcohols delivered the corresponding tetrahydropyrans in high yields and with significant diastereoselectivity. Due to the mild conditions, this process has been applied to the synthesis of numerous natural products. The access to oxepanes has also been considered from 4-pentenols. For example, Padron et al. recently described the synthesis of (+)-isolaurepan **49** by using a catalytic amount of iron(III) and TMSCl as the nucleophile source (Scheme 28). An additional dehalogenation of **68** under radical conditions afforded (+)-isolaurepan in 85 % overall yield [45].

2-Allyl phenol **69** has been similarly implicated into the Prins reaction promoted by stoichiometric amounts of $AlCl_3$ with cyclopentanone providing a straightforward synthesis of benzoxepin **70** (Scheme 29) [46].

Spirocyclic oxindole **73** was prepared from isatin ketal **71** and bis-homoallylic alcohol **72**. The reaction proceeded with a high diastereo- and regioselectivity and with acceptable yield (56 %) (Scheme 30) [47].

2,7-Disubstituted 3-ketooxepane **77** was prepared by condensing cyclopropyl diol **74** with benzaldehyde under Lewis acid catalysis [$Al(OTf)_3$] (Scheme 31). The initially formed ketal **75** smoothly underwent a selective cleavage to produce the oxonium intermediate **76**. Fragmentation of the cyclopropanoate intermediate delivered a nucleophilic species which directly attacked the cationic centre to form **77** (69 %) [48].

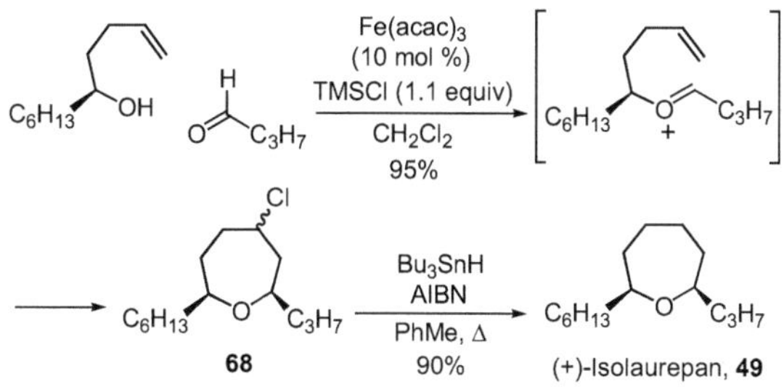

Scheme 28 A two-step synthesis of isolaurepan by Prins reaction and reductive dehalogenation

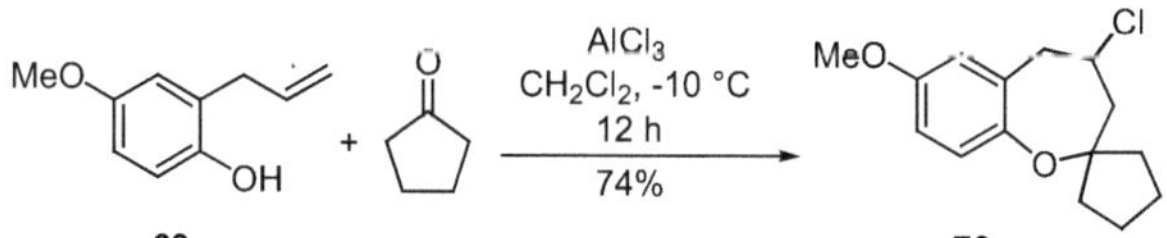

Scheme 29 Formation of spirobenzoxepane from 2-allylphenol and cycloalkanone

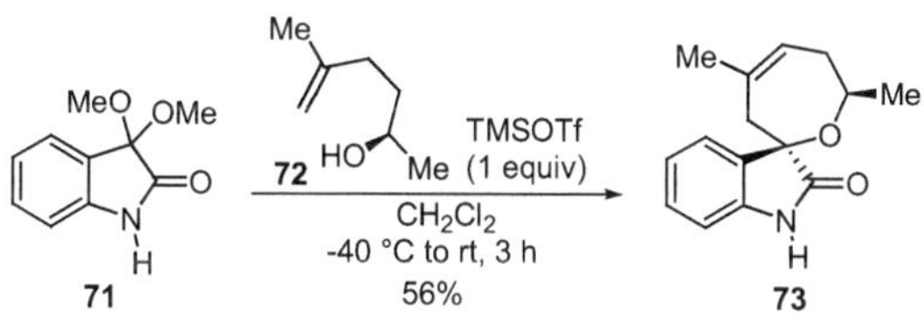

Scheme 30 Prins reaction involving a dimethyl isatin acetal and a bis-homoallylic alcohol

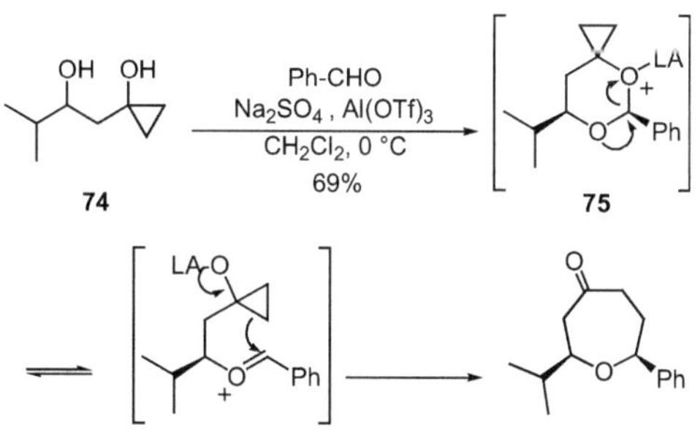

Scheme 31 Synthesis of 4-oxepanone by a tandem ketalisation/ring-opening/ring-closing procedure

Propargylic carbocations can be generated from the corresponding alcohols by complexation of the triple bond with $Co_2(CO)_8$ followed by treatment with a strong Lewis acid (Nicholas-type reaction). When this reaction was applied to ω-epoxy propargylic alcohols, the oxygen atom of the oxirane functionality can act as

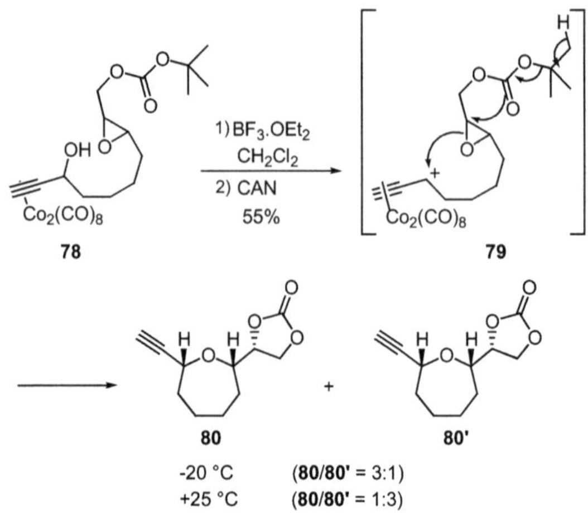

Scheme 32 Synthesis of oxepanes by Nicholas-type reaction

a nucleophile to form various-sized heterocycles. Applied to **78**, the nature of the protective group of the primary alcohol was particularly crucial; only the *tert*-butylcarbonate derivative led to a mixture of oxepanes **80** and **80′**. Interestingly, a strong temperature effect was also noticed on the diastereomeric ratio. At −20 °C, the ratio *syn*-**80**/*anti*-**80′** was 3/1, and at +25 °C, the ratio was reversed to 1/3 (Scheme 32) [49].

4.2 *Anionic Cyclisations*

4.2.1 Intramolecular Cyclisation of ω-Tributylstannyl Ether Aldehydes

Initially developed by Yamamoto et al. [50], the formation of alkoxyallylstannanes and their further intramolecular addition on aldehydes has been extensively used to access complex marine polycyclic ethers such as (−)-brevisin [51] (Fig 2). When compound **81** was treated with methoxyallylstannane **82**, the transient alkoxyallylstannane **83** was generated and transformed to **84** in a very good yield (Scheme 33). This approach has been also investigated to access other polycyclic ethers as hemibrevetoxin B. In this latter case, the formation of the vinyl group was combined with a ring-closure metathesis to achieve the formation of a subunit containing two fused oxepanes [52].

4.2.2 Intramolecular Wittig Reaction and Aldolisation

Few methods were published regarding the formation of oxepanes under basic conditions. Oxepanone **86**, precursor of zoapatanol, has been obtained by an

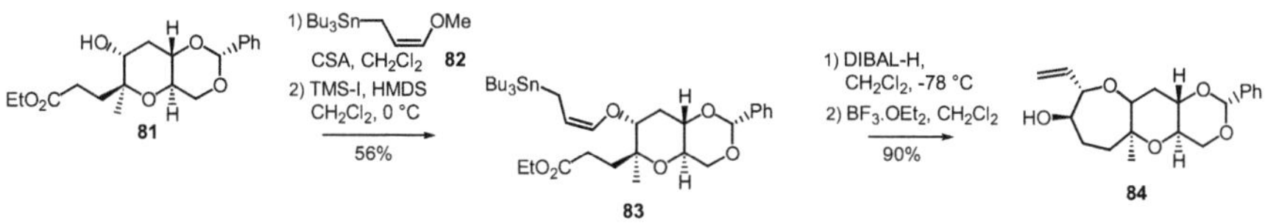

Scheme 33 Yamamoto's approach to hemibrevetoxin B

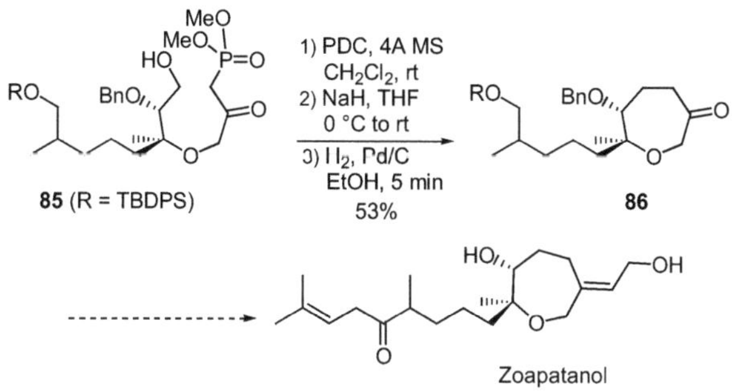

Scheme 34 Sequential oxidation/Wittig–Horner reaction and hydrogenation: access to a zoapatanol precursor

intramolecular Wittig–Horner reaction followed by the hydrogenation of the newly created double bond (Scheme 34) [53].

Bach et al. have achieved the total synthesis of punctaporonin C based on a highly stereoselective [2+2]-photocycloaddition [54]. Due to the shape of the tricyclic photoadduct **87**, an intramolecular aldolisation allowed the formation of the additional seven-membered ring in 56 % yield. A β-elimination led to an enone which was finally reduced by a catalytic hydrogenation in the presence of Crabtree catalyst (Scheme 35).

In contrast with the two previously mentioned syntheses based on anionic condensations, the synthesis of pterulone has required an intramolecular nucleophilic substitution of an allylic chloride by a sulphone. After optimisation of the reaction conditions, the access to the core structure of the natural product, by using LHMDS as base and in the presence of LiBr as additive, has been successful as **91** was isolated in 78 % yield (Scheme 36) [55].

4.2.3 Brook Rearrangement/Conjugate Addition

For their astonishing six-step synthesis of strychnine **94**, Martin and Vanderwal described an unprecedented sila-Brook rearrangement followed by a conjugate addition of a transient vinyl cuprate to an alkenal. The overall yield of the process was 5–10 %. This apparent low yield can be counterbalanced by the access to the D

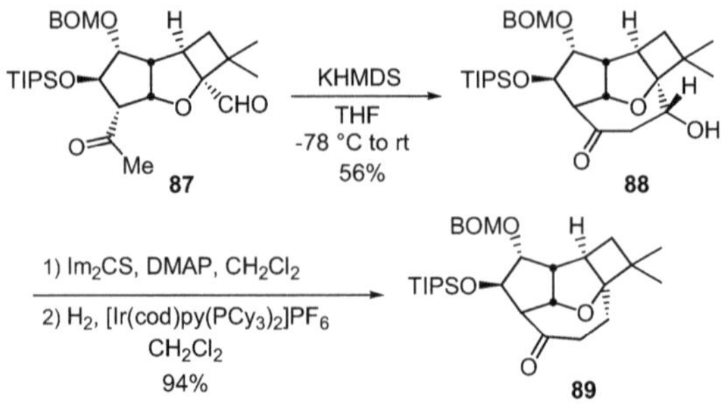

Scheme 35 Intramolecular aldolisation in the synthesis of punctaporonin C

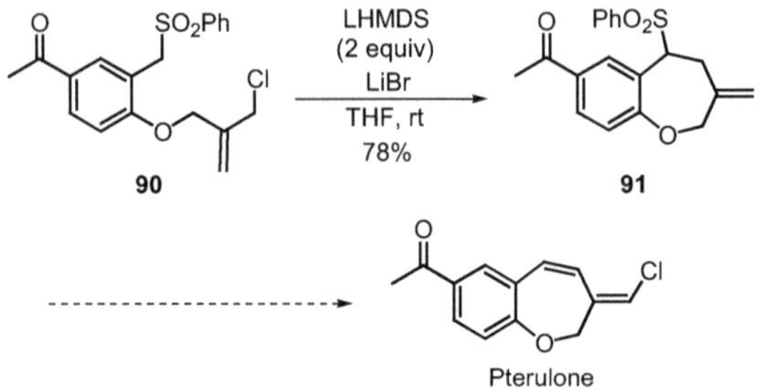

Scheme 36 Access to the pterulone core structure by intramolecular substitution

and F rings of strychnine in one single step. In addition, intermediate **93** was transformed in one single step to strychnine **94** (Scheme 37) [56].

4.3 Metal-Induced Cyclisations

4.3.1 Palladium-Promoted Cyclisation

Palladium plays a crucial role in organic synthesis for the formation of single or multiple C–C bonds from aryl derivatives. Intramolecular Heck reaction has been particularly useful to access seven-membered rings. Nielsen et al. have studied the regioselectivity of these reactions from unsubstituted substrates and found that the regioselectivity depends on the nature of phosphine ligands utilised [57]. A moderate selectivity in favour of the oxepin **97** was thus observed with S-Phos (Scheme 38).

The Heck reaction has been particularly investigated to access pharmaceutically active compounds such as olopatadine hydrochloride, an anti-allergic drug (Scheme 39) [58, 59]. When **98** was treated with Pd(0) in the presence of formic acid and piperidine, oxepin **99**, precursor of olopatadine, was isolated in 47 % yield.

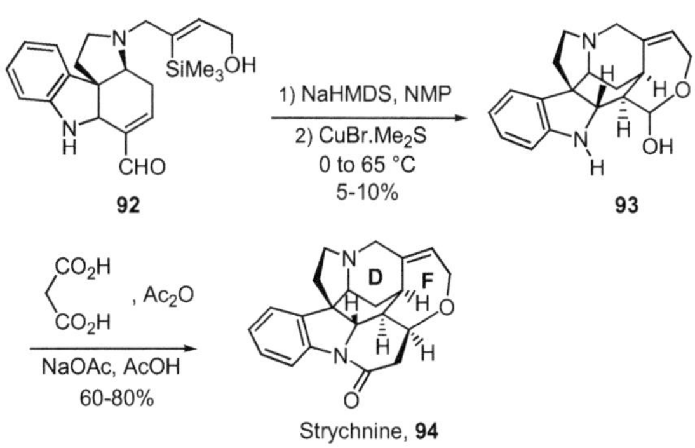

Scheme 37 Brook rearrangement and Michael addition for the simultaneous formation of the D and F rings of strychnine

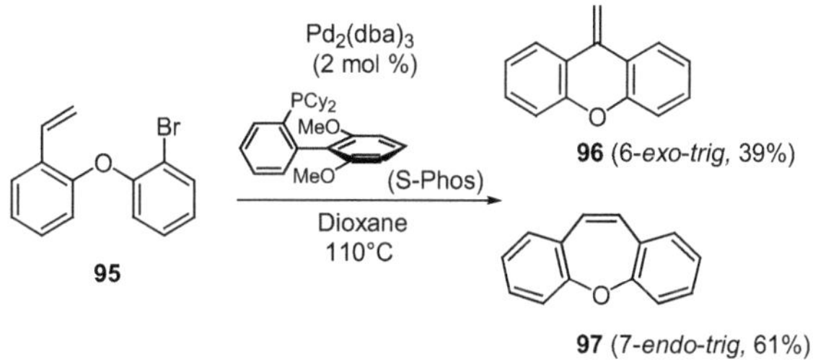

Scheme 38 Access to dibenzooxepins by an intramolecular Heck reaction

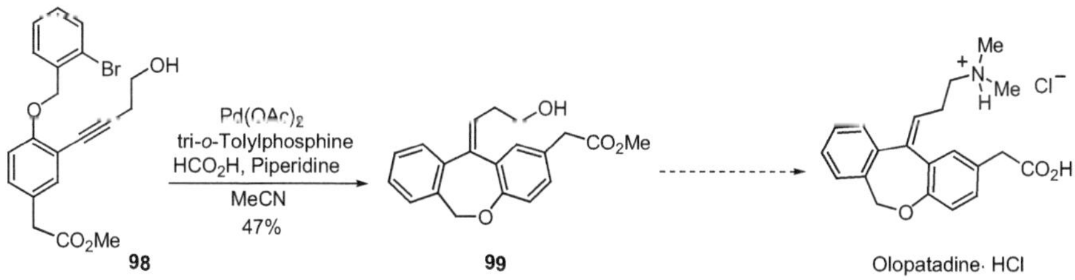

Scheme 39 Application of an intramolecular Heck reaction to the synthesis of olopatadine

The intramolecular coupling between a vinyl iodide and an allylic alcohol under Jeffery's conditions was advantageously applied to a 12-step total synthesis of strychnine **94** by MacMillan et al. (Scheme 40) [60].

A few years ago, Lautens et al. have developed a straightforward synthesis of benzoxepines by combining a Catellani-type palladium-catalysed aromatic substitution with an intramolecular Heck reaction to form **104** from **102** to **103** (Scheme 41) [61].

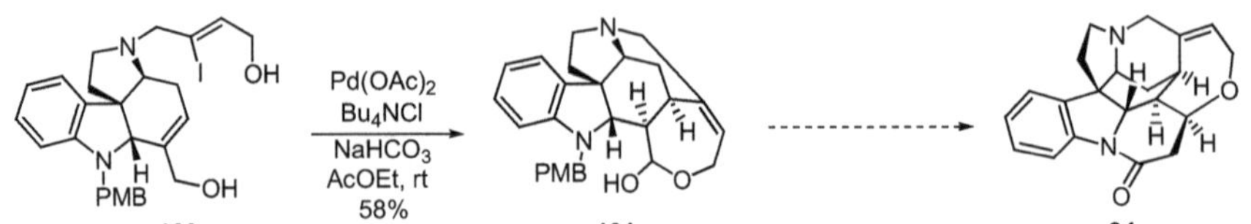

100 **101** **94**

Scheme 40 Intramolecular Heck reaction/ketalisation used in MacMillan's strychnine synthesis

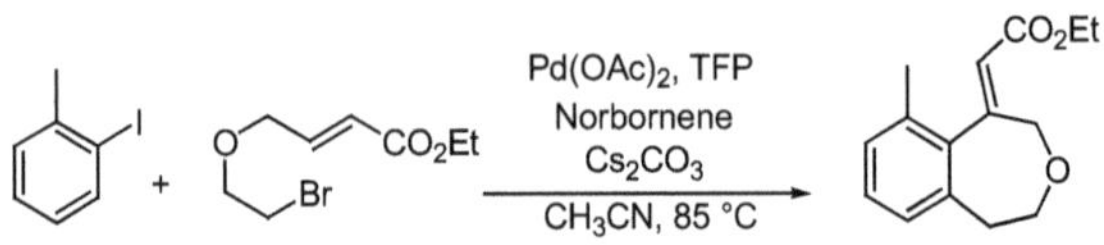

102 **103** **104**

Scheme 41 Formation of benzoxepanes by palladium-catalysed cyclisation

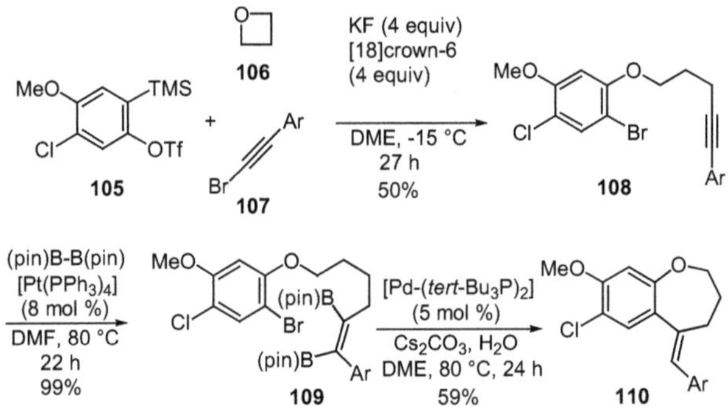

105 **106** **107** **108** **109** **110**

Scheme 42 Synthesis of benzo[*b*]oxepin by diborylation/intramolecular Suzuki coupling

In a similar context, a three-component coupling of an in situ-generated benzyne from **105**, oxetane **106** and alkynyl bromide **107** delivered the corresponding *o*-bromo pentynyl phenol ether **108** in good yield (Scheme 42). This substrate could easily undergo a platinum-catalysed diborylation which was followed by a Suzuki–Miyaura coupling. Protodeborylation achieved under basic conditions afforded the benzo[*b*]oxepin **110** [62].

An alternative procedure has been more recently designed which combines a palladium-catalysed bis-cyclisation from iodoalkyne **111** and benzyne precursor **112**, leading to the polycyclic structure **113** (Scheme 43) [63].

4.3.2 Gold Catalysis

Intramolecular gold-catalysed hydrofunctionalisation of alkynes and allenes has emerged as a valuable method to oxepanes. The nature of the catalyst and the

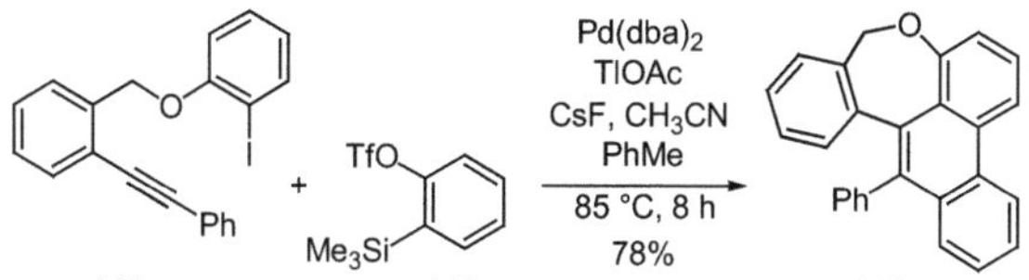

Scheme 43 Synthesis of polyaromatic structures by a palladium-catalysed bis-cyclisation

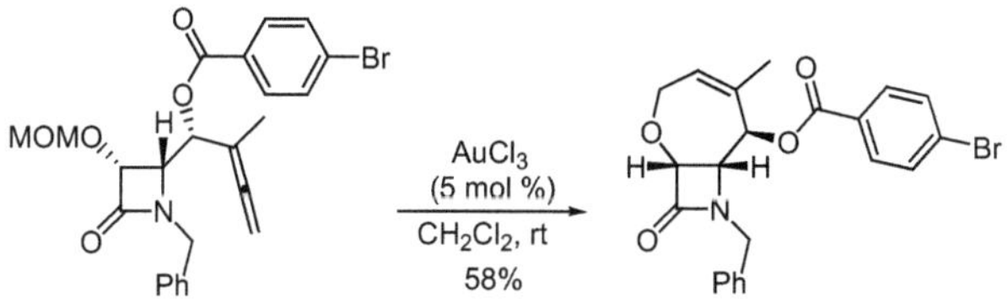

Scheme 44 Cyclisation of hydroxyallene **114** to tetrahydroxepin **115** under gold catalysis

protective group fixed on the nucleophilic species plays a crucial role on the regioselectivity of the reaction. The work of Alcaide and Almendros et al. illustrates the power and the efficiency of such process and has been demonstrated in the synthesis of **115** from **114** using $AuCl_3$ as catalyst (Scheme 44) [64].

The formation of the seven-membered ring was explained by a selective attack of the oxygen atom of the methoxymethyl ether group of **114** onto the terminal position of the allene gold complex. Protonolysis afforded the tetrahydrooxepine framework with concomitant liberation of methanol and formaldehyde.

4.3.3 Metathesis

Since the discovery of air-stable catalysts which can be used with highly substituted substrates, metathesis reactions have revolutionised the disconnection approach to complex molecules. Largely used for the formation of five- and six-membered rings or even very large macrocycles, the access to oxepins and benzoxepines has been far less considered [65]. The successful syntheses of different natural products containing oxepanes and oxepenes from readily available substrates have been reported.

Total synthesis of rogioloxepane A was thus completed in 21 steps by Crimmins and DeBaillie (Scheme 45). The oxepene ring was obtained by a ring-closing metathesis (RCM) performed on **116** with Grubbs first-generation catalyst with an outstanding 96 % yield [66].

Eranthin **119**, an oxepinochromone natural product, has been prepared from diene **118** by an RCM process. The formation of the seven-membered ring occurred with a very high yield (84 %) at the final stage of the synthesis (Scheme 46) [67].

Related benzodioxepanes and indanonyl oxepanes have been prepared by using an RCM as the key step [68, 69]. Janoxepin, a compound which displays

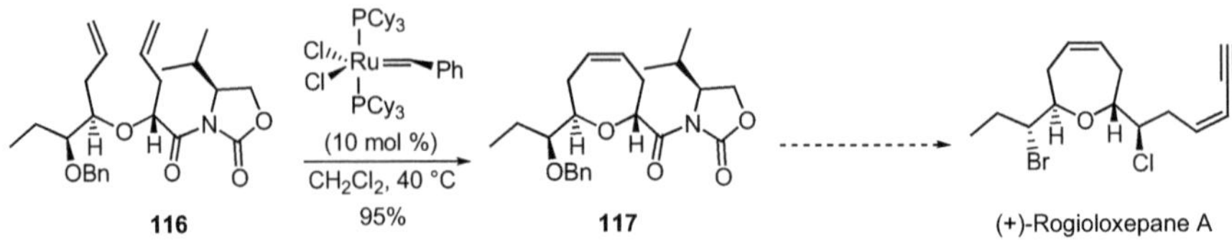

Scheme 45 RCM leading to the core structure of (+)-rogioloxepane

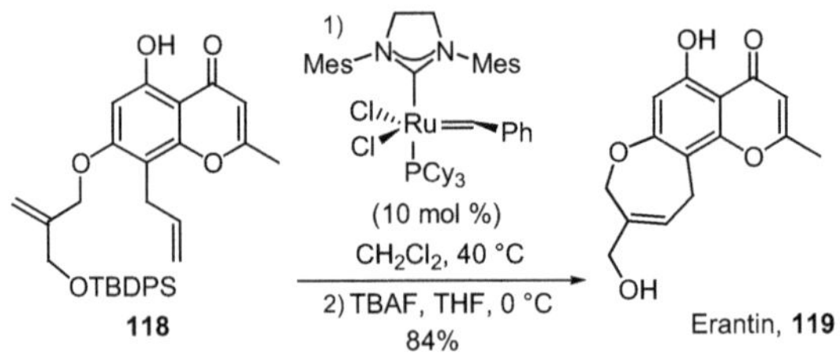

Scheme 46 Application of RCM to the synthesis of eranthin

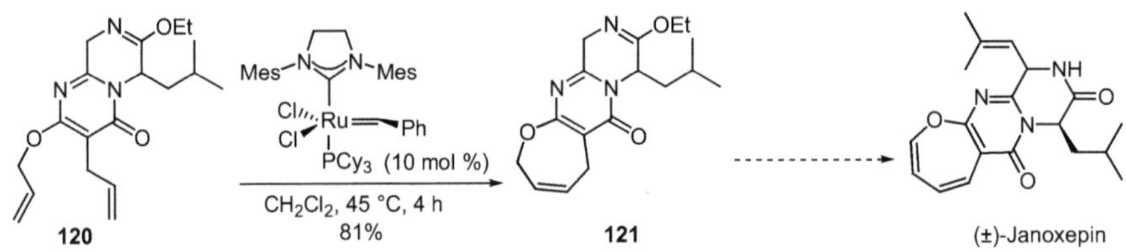

Scheme 47 RCM en route to (±)-janoxepin

antiplasmodial activity against *Plasmodium falciparum*, has been prepared from diallyl pyrimidinone **120**. After protection of the lactam moiety, the RCM delivered the oxepene **121** in 81 %, and this compound was then converted into the expected janoxepin (Scheme 47) [70].

In the context of a total synthesis of aranotin, Bräse et al. have described an efficient access to the oxepin subunit by applying an RCM to an ω-enol ether alkene. Interestingly, the reaction was successful only after the removal of the protecting group of the allylic alcohol (Scheme 48) [71].

Iridium-catalysed etherification of *o*-allylphenol **125** with unsaturated carbonate **126** afforded a diene with high enantiomeric excess. In the presence of Zhan's catalyst **I** (2 mol%), the RCM furnished the 2.5-dihydrobenzo-[*b*]oxepin **127** with the same level of enantiomeric excess (Scheme 49) [72].

It is noteworthy that during the RCM, the migration of the double bond can occur, due to the in situ generation of a ruthenium-hydride species. Schmidt et al. have reported a short access to two isomeric oxepenes **129** and **130** starting from the same material **128** and using the same Grubbs catalyst (**GI**) but under two different conditions (Scheme 50) [73].

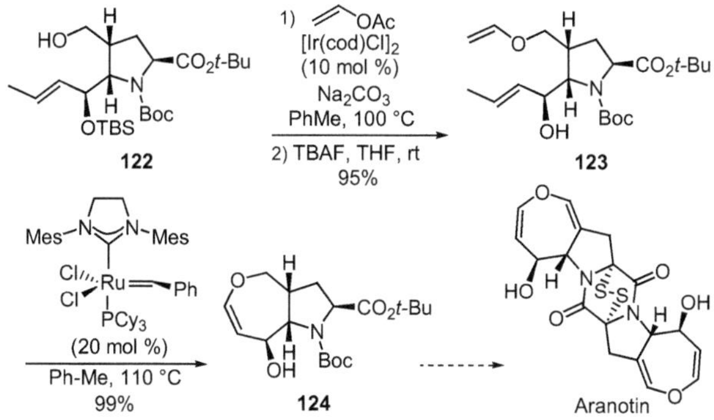

Scheme 48 RCM between an enol ether and an allylic alcohol to reach the oxepin ring of aranotin

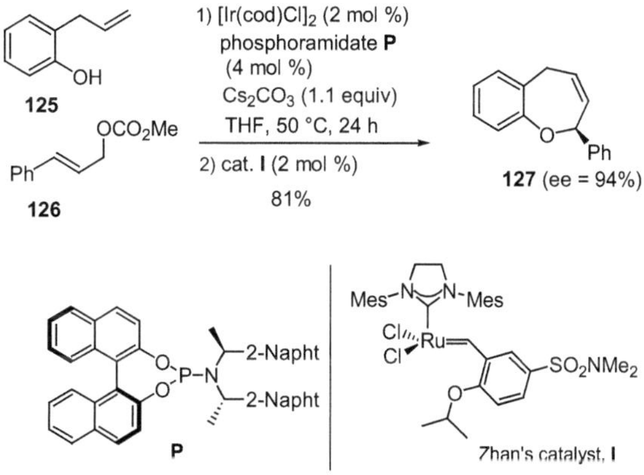

Scheme 49 Combination of iridium and ruthenium catalysis to access enantioenriched benzoxepine **127**

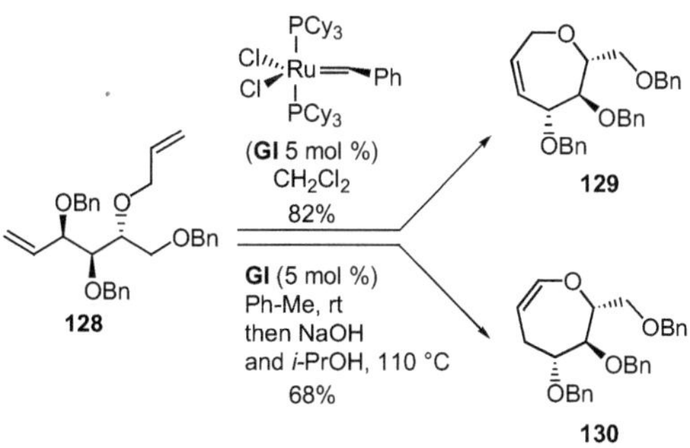

Scheme 50 Complementary approach to oxepene **129** or **130** from the same diene **128**

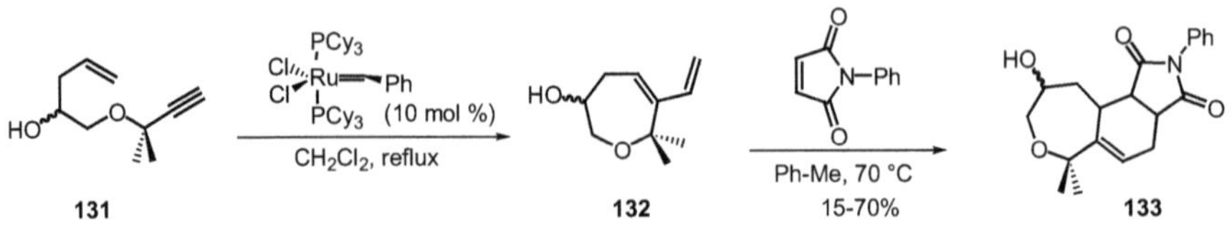

Scheme 51 Sequential ene-yne metathesis/Diels–Alder cycloaddition

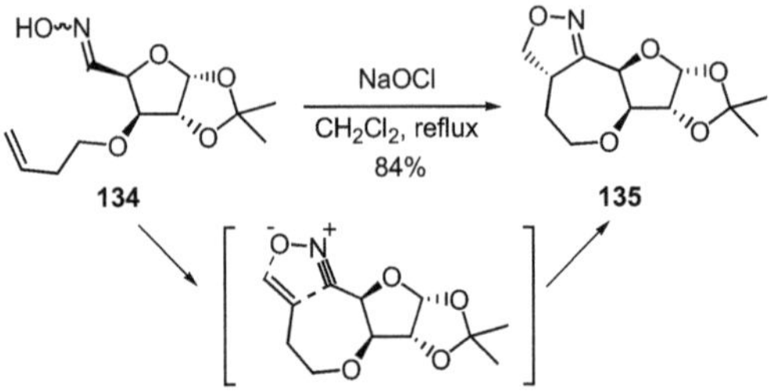

Scheme 52 Formation of isoxazolinoxopane **135** by a diastereoselective [3+2]-cycloaddition

Compared to RCM of 1,8-dienes, ene-yne metathesis has been more scarcely used. However, **131** was transformed to vinyl oxepene **132** which was directly involved in a Diels–Alder reaction to deliver a new library of tricyclic compounds with potent protein kinase inhibition properties (Scheme 51) [74].

4.4 *Thermal Cycloadditions and Rearrangements*

O-Homoallyl nitrile oxide **134** readily prepared from sugars has been involved in an intramolecular nitrile oxide/alkene cycloaddition (INOAC) to deliver fused isoxazolinoxopane **135**. The reaction proceeded according to an *exo*-mode with a high diastereoselectivity (Scheme 52) [75, 76].

A Cope-type [3,3]-sigmatropic rearrangement was observed when cyclopropylmethyl alcohol **136** was oxidised under Swern conditions. The irreversible formation of the cyclic ether **138** from intermediate **137** was probably due to the release of the cyclic strain and also due to the formation of a stabilised silyl enol ether (Scheme 53) [77].

The electrocyclic ring opening of bicyclo[2.1.0]oxaheptane **140** prepared by epoxidation of cyclobutene derivative has been carried out at high temperature. The reaction performed in the presence of 2,6-*tert*-butyl-4-methyl-phenol (BHT) to prevent extensive polymerisation afforded a mixture of two stereoisomers **141** and **142** without significant selectivities (Scheme 54). This low value can be attributed to the formation of a biradical intermediate [78].

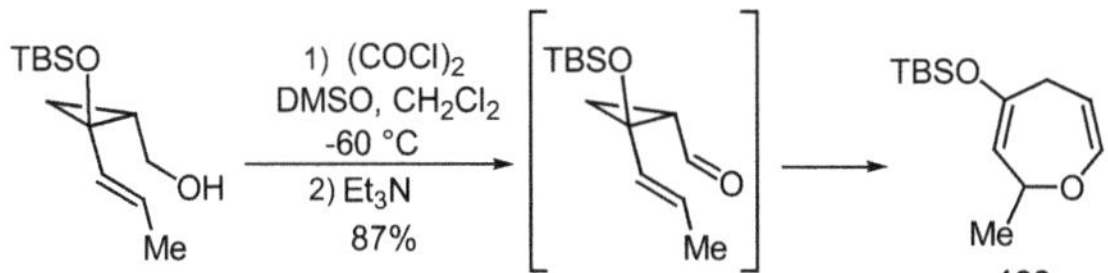

136 137 138

Scheme 53 Access to a seven-membered ring ether involving a Cope rearrangement

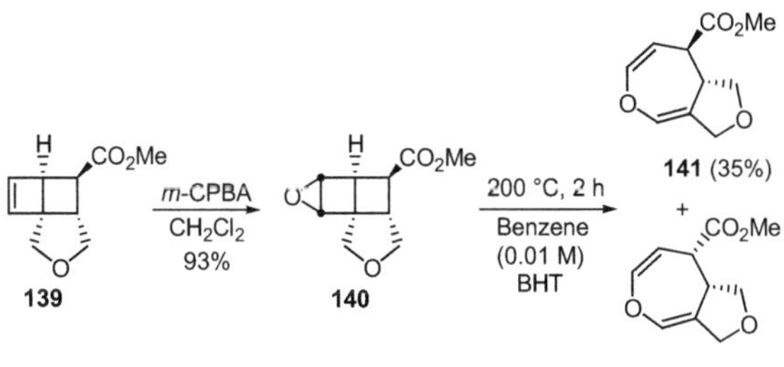

139 140 **141** (35%) **142** (46%)

Scheme 54 Thermal electrocyclic ring-opening of bicyclo[2.1.0]oxaheptane **140**

4.5 Radical Reactions

4.5.1 Free Radical Cyclisation

Nowadays, radical cyclisation is still an ongoing domain. Instead of the standard anionic and cationic processes, the radical reactions can be carried out on a wide array of substrates without extensive use of protecting groups. Nevertheless, the selective formation of seven membered rings is not a favourable process according to Baldwin's rules, and hydrogen atom abstraction usually takes place, decreasing the efficiency and attractiveness for this radical process.

Regarding the total synthesis of ciguatoxin (Scheme 55), the G oxepane ring was constructed from **143** according to a 7-*exo*-trig process leading to a secondary radical, stabilised by the presence of an electron-withdrawing group. To avoid extensive oligomerisation, the reaction was conducted under highly diluted conditions with complete stereocontrol and with an impressive 85 % yield [79].

7-*Endo*-trig cyclisations have been also considered to prepare new topoisomerase inhibitors from 3-aryl-isoquinolines (Scheme 56) [80]. Thus, in the case of **146**, the high regioselectivity allowing the formation of **148** could be easily explained by the stability of the intermediate (secondary radical α to an oxygen atom) before the reduction with tributylstannyl hydride.

A fruitful 1,2-radical rearrangement has been also reported to convert 2-(9*H*-xanthenyl)malonate monoester **149** into dibenzo[*b*]oxepin **150**. The process required the use of manganese(III) triacetate as a single-electron transfer (SET) oxidant. The primary and highly stabilised radical could then undergo a ring

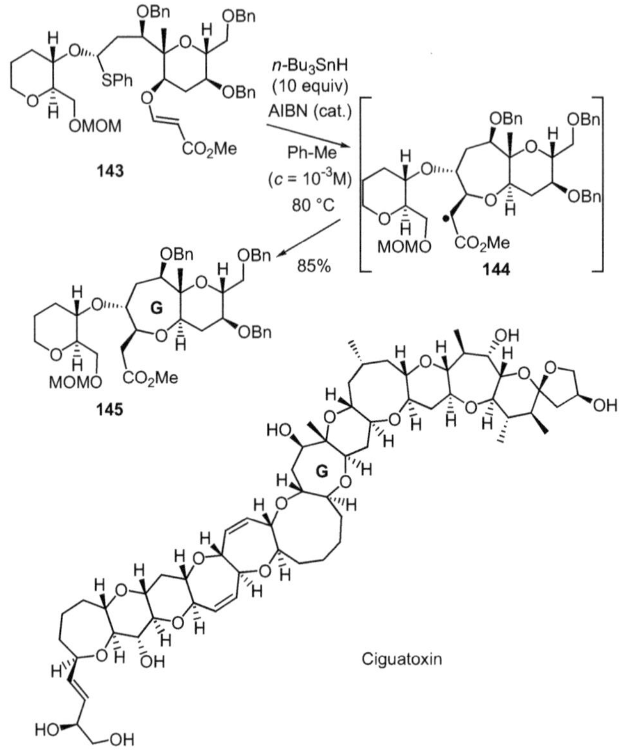

Scheme 55 Radical cyclisation used to generate the G ring of polyether ciguatoxin

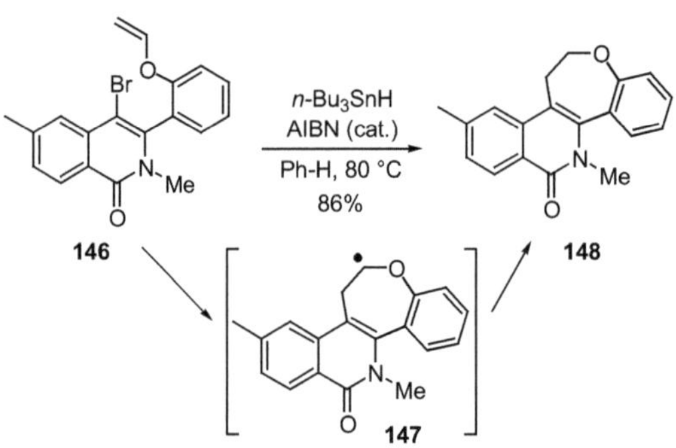

Scheme 56 7-*Endo*-trig cyclisation leading to the formation of a tetracyclic oxepin

expansion leading to a new benzylic radical intermediate. A concomitant oxidative decarboxylation/elimination promoted by a second equivalent of manganese salt could then deliver dibenzooxepin **150** (Scheme 57) [81].

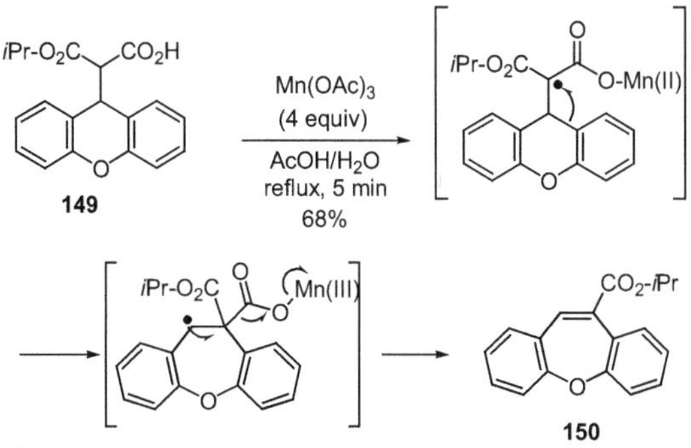

Scheme 57 Ring enlargement of a xanthene derivative into dibenzo[*b*]oxepin

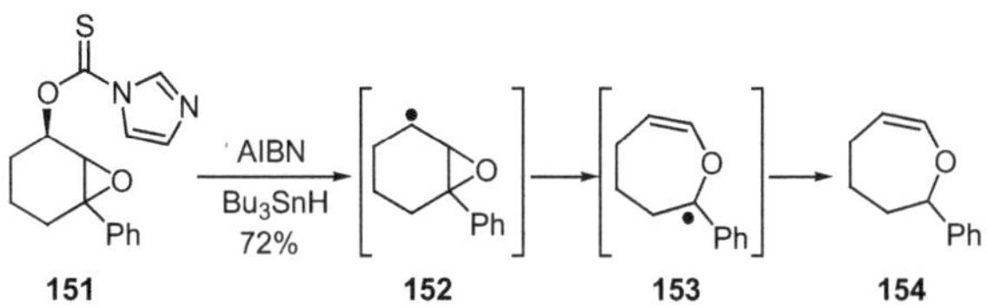

Scheme 58 Ring enlargement of cyclohexene oxide under radical conditions

4.5.2 Radical Ring Enlargement of Cyclohexene Oxides

Taken advantage of the easily homolytic cleavage of thiocarbonate derivatives, the ring opening of cyclohexene oxide in **151** delivered the corresponding oxepene **154** in good yield. No doubt that the presence of the phenyl group has a significative influence on the course of the reaction to preferentially deliver the benzylic radical **153** (Scheme 58) [82, 83].

4.5.3 Samarium-Induced Cyclisations

Samarium diiodide is a unique reagent developed by Kagan et al. and later by Molander et al. to promote C–C bond formation and C–C bond cleavage. In the context of natural product syntheses, SmI_2 is of great interest, as this reagent is compatible with numerous functionalities such as esters, unprotected alcohols. Nakata et al. reported the formation of the F and H rings of gambierol in a single step by transformation of **155–156**. The unique stereochemistry observed in **156** was rationalised by a transition state in which the oxygen atom of the ketyl group and the ester group is chelated by samarium (Scheme 59) [84]. This high-yielding method has been used iteratively to build other oxygenated rings in a number of molecules with similar efficiency [11, 85].

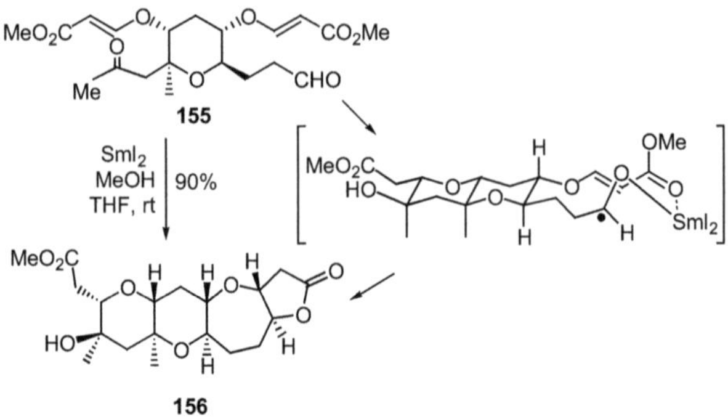

Scheme 59 Simultaneous formation of the F and H rings of gambierol promoted by samarium diiodide

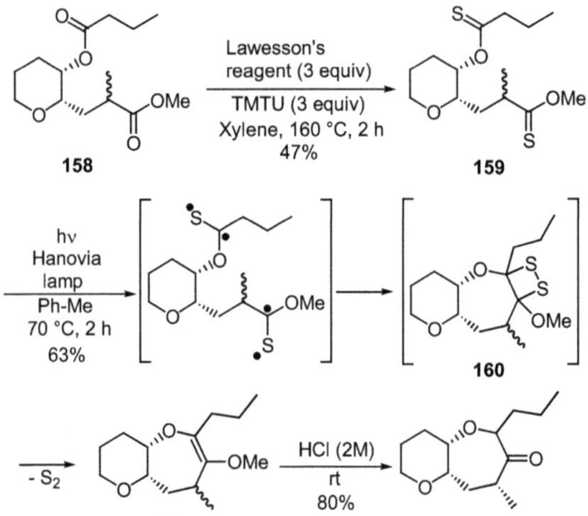

Scheme 60 Access to oxopanone **162** through the formation and cleavage of a 1,2-dithietane generated under UV irradiation

4.6 *Photochemical Processes*

In the course of the total synthesis of brevetoxin B, Nicolaou et al. have designed a new and elegant access to oxepenes from dithiono derivatives. Thus, irradiation of **159** can generate a bi-diradical intermediate which can furnish after recombination of the bicyclic 1,2-dithietane **160**. Under UV irradiation, this unstable structure can lose sulphur leading to the formation of a C=C bond (Scheme 60). The initially attempted transannular process was unsuccessful due to a problematic dithionation of bis-lactone precursors. Therefore, the reaction was efficiently developed on a large number of acyclic structures related to the complex target molecule (Scheme 61) [32].

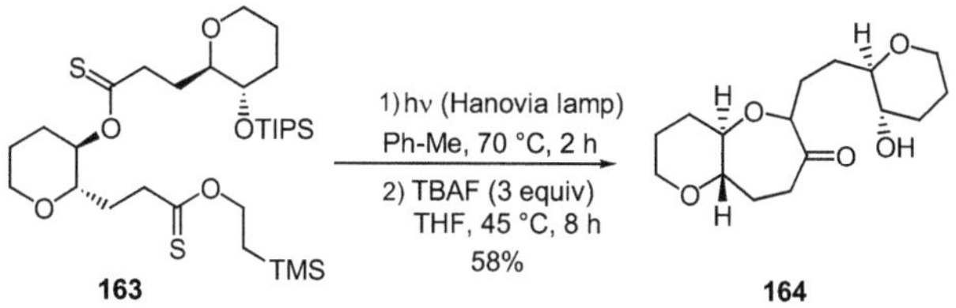

Scheme 61 Access to oxopanone **164** by irradiation of bis-thioester **163**

5 Access to Seven-Membered Ring Lactones

A large number of methods are to the disposal of chemists to achieve the synthesis of medium-sized lactones. Some of them, applied to the synthesis of caprolactones, are depicted vide infra.

5.1 Oxidative Processes

5.1.1 Oxidation of Diols/Lactols and Lactonisation

The oxidation of 1,6-hexanediol is a very important chemical process, as it gives a short access to caprolactone, a major compound in polymer chemistry. This has been done by using a wide panel of oxidising reagents. Eco-friendly methods combining hydrogen peroxide and heteropolyacid catalysts are very appealing for this purpose [86]. Noteworthy also is the ruthenium-catalysed aerobic oxidation of diols which requires highly diluted conditions to prevent extensive oligomerisation [87]. Another oxidative lactonisation based on hydrogen transfer, catalysed by ruthenium salts (1 mol%) in acetone, has also been tested. While the reaction was inefficient with linear hexan-1,6-diol, introduction of a rigid biphenyl backbone led, nearly quantitatively, to the expected lactone (Scheme 62) [88].

Access to functionalised lactones by selective oxidation of unsymmetrical diols has also been considered. In the context of the total synthesis of (−)-brevenal, a complex natural product possessing five oxygenated fused rings, the selective oxidation of the primary alcohol in **169** followed by a lactonisation and a subsequent oxidation furnished in a one-pot process, the expected lactone **170** [89]. While primary alcohols react more rapidly than secondary ones, the overall yield is usually excellent, as demonstrated by the conversion of **171–172** in the course of the total synthesis of the even more challenging maitotoxin (Scheme 63) [90].

5.1.2 Baeyer–Villiger Reaction

The ring enlargement of cyclohexanones has been considered to prepare homobrassinolides which possess potential neuroprotective activities. Of interest, the

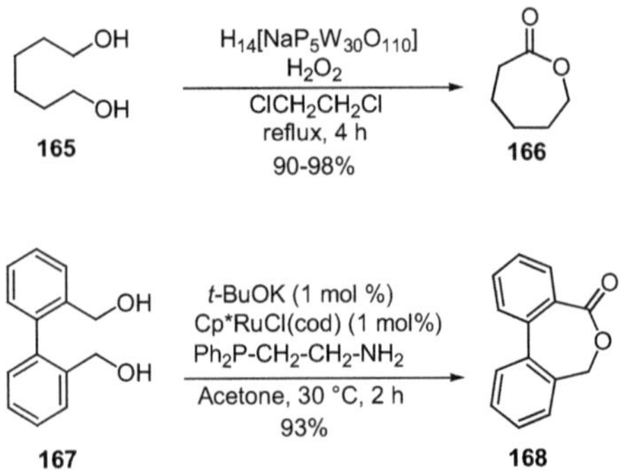

Scheme 62 Direct formation of caprolactone by oxidation of a symmetrical diol

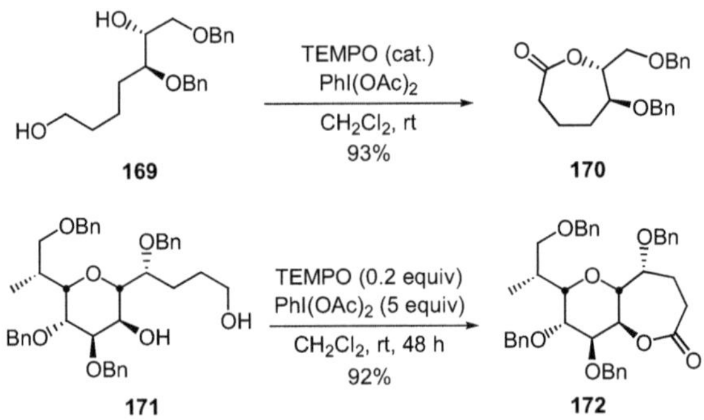

Scheme 63 Access to caprolactones by oxidation of diols with TEMPO/PhI(OAc)$_2$

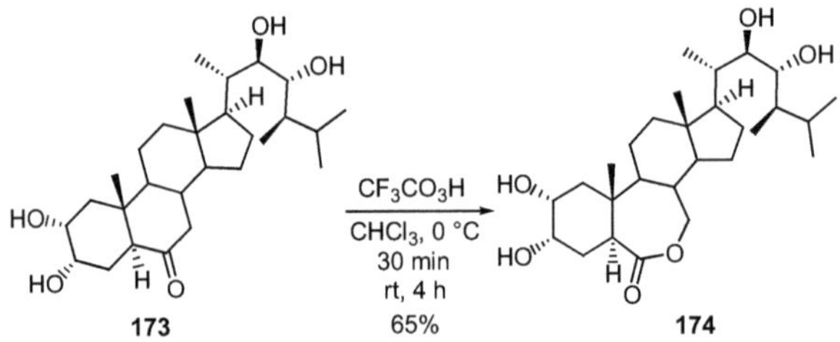

Scheme 64 Application of the Baeyer–Villiger reaction to the ring expansion of brassinolide **173**

reaction was performed without protection of tetrol **173** when treated with CF_3CO_3H delivering regioselectively the seven-membered ring lactone **174** in 65 % yield (Scheme 64) [91].

A key intermediate for the synthesis of the core structure of platensimycin, an unusual and very promising antibacterial agent, was prepared through a similar way. When **175** was treated with *m*-CPBA, the introduction of the oxygen atom occurred in the C1–C2 bond to produce **176** in 73 % (Scheme 65) [92].

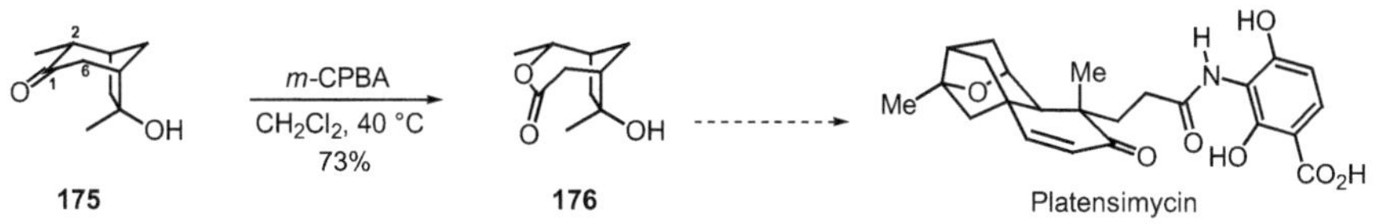

Scheme 65 Baeyer–Villiger reaction applied to a bicyclic ketone

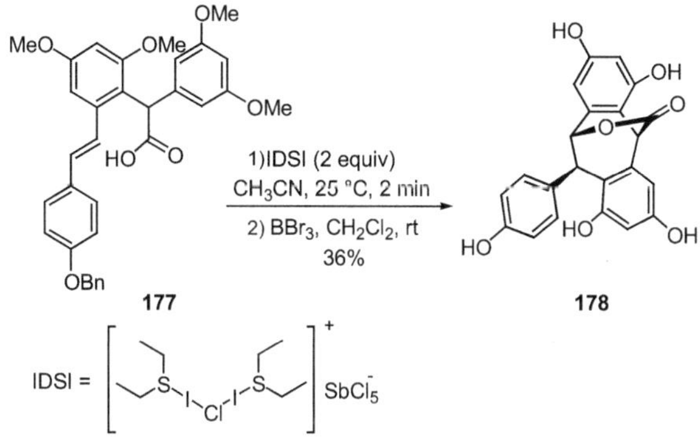

Scheme 66 Cascade halolactonisation/Friedel–Crafts reaction promoted by IDSI reagent

5.2 *Halolactonisation*

A cascade halolactonisation/Friedel–Crafts reaction has been investigated during the total synthesis of polyphenol derivatives. The transformation of **177–178** required the use of the recent developed iodo diethyl sulfonium iodopentachloroantimonate (IDSI) reagent [93] (Scheme 66). We have to point out that other halonium sources led to the decomposition of the starting material [94].

5.3 *Lactonisation of ω-Hydroxy Acids*

Esterification and macrolactonisation can be performed from hydroxy acids by the in situ activation of a carboxylic acid [95]. Yamaguchi's conditions have been widely used to access medium-sized rings. This reaction represents the key step in the synthesis of luffalactone by Basabe et al. and allowed the transformation of **179–180** in 88 % yield (Scheme 67) [96]. Similarly, Fuwa and Sasaki took advantage of this method to build the E ring of gambieric acid during their efforts toward its total synthesis (Scheme 68) [97].

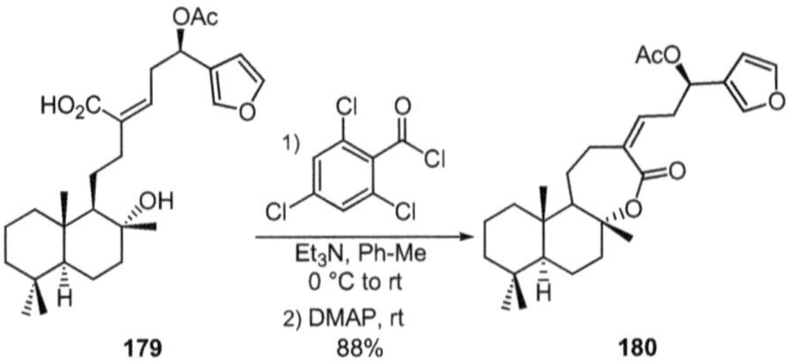

Scheme 67 Formation of caprolactone **180** under Yamaguchi's conditions

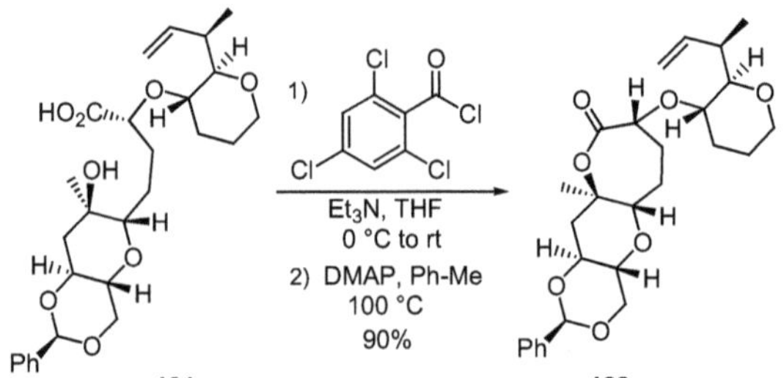

Scheme 68 Lactonisation of hydroxy acid **181** under Yamaguchi's conditions

5.4 *Tandem Suzuki Coupling/Macrolactonisation*

A tandem reaction has been devised allowing a direct access to various resorcylic lactones, a class of natural products possessing various alkoxy groups on the aryl moieties. The tandem Suzuki coupling/macrolactonisation sequence was applied to **183** and **184** to furnish, after hydrogenation, graphislactone D, **185** (Scheme 69) [98, 99].

5.5 *Ring Enlargement*

5.5.1 Thermal [2+2]-Cycloaddition/Electrocyclic Ring Opening

α,β-Unsaturated caprolactone **189** has been prepared from 2-methoxy-tetrahydrofuran **186** and 1-(trisisopropylsilyloxy)alkyne **187**. The reaction, performed at 0 °C in CH_2Cl_2, required a Lewis acid activation (e.g. $BF_3{\cdot}OEt_2$). In a mechanistic point of view, the oxocarbenium **186′** was generated in situ; the resulting trifluoro(alkoxyborate) reacted with the silyoxyalkyne to deliver ynolate **187′**. A stepwise reaction between these two highly reactive species led to an oxetenium intermediate **188** which immediately underwent an electrocyclic

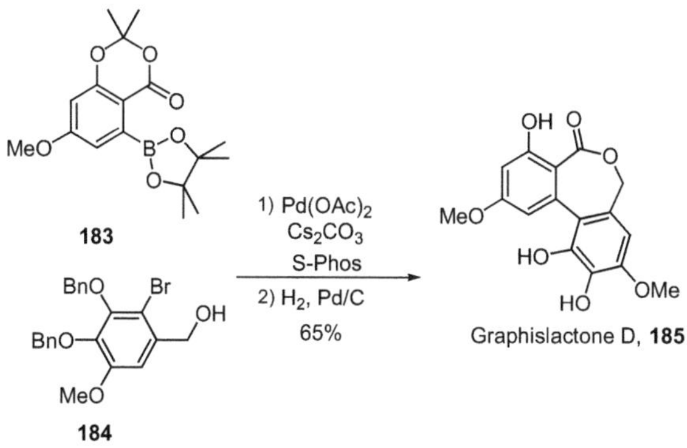

Scheme 69 Access to graphislactone D by a tandem Suzuki coupling/macrolactonisation

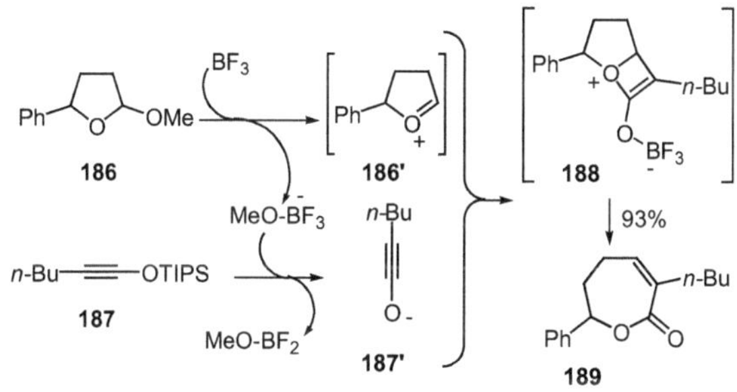

Scheme 70 Access to an unsaturated lactone by an electrocyclic ring opening of an oxetenium intermediate

ring opening to deliver the unsaturated seven-membered ring lactone in high yields (Scheme 70) [100].

5.5.2 Ring Expansion of Oxacyclohexane-2-Carboxaldehydes

The organo-catalysed rearrangement of oxacycloalkane-2-carboxaldehydes has been successfully achieved by treatment with *N*-heterocyclic carbene (NHC) under basic conditions. While the reaction proceeded well with furan derivatives, the extension to the synthesis of seven-membered ring lactones was more problematic in terms of time and yields. For example, treatment of **190** with carbene **191** (10 mol%) led to lactone **192** (48 %) after 10 days (Scheme 71) [101].

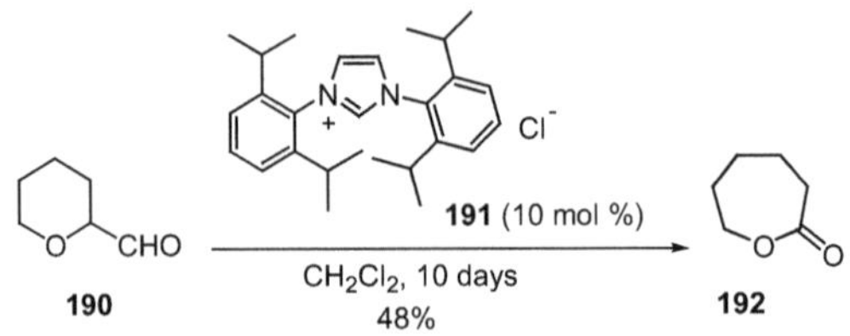

Scheme 71 Rearrangement of oxacyclohexane-2-carboxaldehyde promoted by NHC

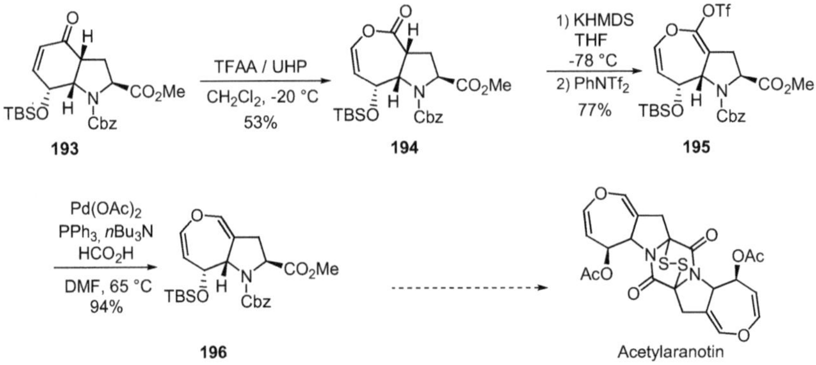

Scheme 72 Tokuyama's synthesis of acetylaranotin

6 From Lactones to Oxepins and Oxepenes

As disclosed above, several methods have been published to access medium-sized ring lactones. These substrates can also be transformed to oxepins by reduction of transient enol ethers. During the total synthesis of (−)-acetylaranotin, Tokuyama et al. reported a three-step procedure to access the key intermediate **196** from hexahydroindolone **193**. Regioselective Baeyer–Villiger reaction using urea hydroperoxide (UHP) as the oxidant delivered lactone **194** which was efficiently converted to enol ether **195**. Under palladium-catalysed reduction using formic acid as the hydride donor, compound **196** was isolated in excellent yield (Scheme 72) [102].

Enol phosphates can alternatively be implicated into coupling reactions with organocopper reagents. A total synthesis of (±)-isolaurepan **39** took advantage of this strategy. The cyclic enol phosphate **198** was immediately reduced with triethylsilane to deliver oxepane **39** in 74 % yield as a unique *cis*-stereoisomer (Scheme 73) [89].

Other organometallic coupling processes have been investigated. For example, Suzuki–Miyaura coupling between enol phosphate **201** generated from lactone **200** and alkylborate **202** led to the formation of oxopene **203** in 90 % yield (Scheme 74). This compound represents a key fragment for the synthesis of (−)-brevenal [103].

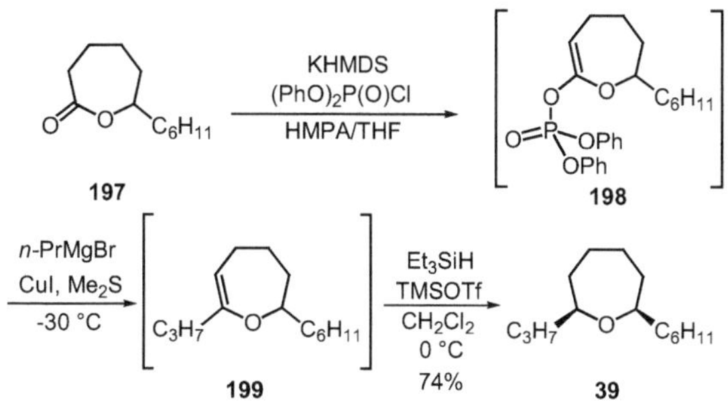

Scheme 73 Two-step total synthesis of isolaurepan **39** from lactone **197**

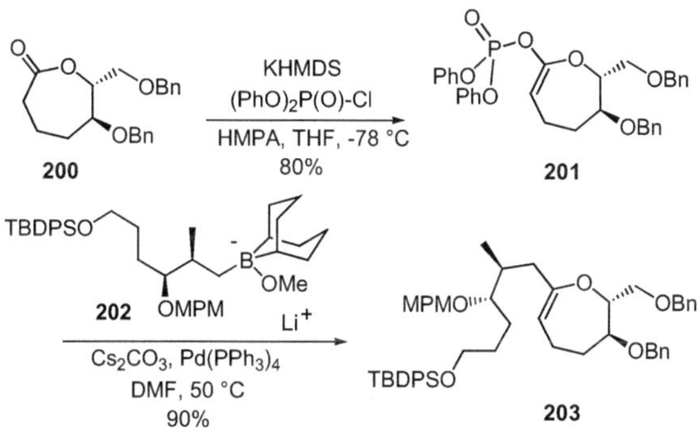

Scheme 74 Suzuki–Miyaura coupling achieved from enol phosphate of caprolactone **200**

A similar procedure was also carried out for the synthesis of fragment ABCDEFG of maitotoxin [90].

7 Conclusion

A great variety of methods are nowadays to the disposal of organic chemists for synthesising oxygenated seven-membered rings. Among them, catalytic processes (RCM and palladium-catalysed cross-coupling) are the more attractive. There is no doubt that even more efficient procedures will be discovered in the near future and applied to the total synthesis of complex natural products.

References

1. Hoberg JO (1998) Tetrahedron 54(42):12631–12670
2. Snyder NL, Haines HM, Peczuh MW (2006) Tetrahedron 62(40):9301–9320

3. Kleinke AS, Webb D, Jamison TF (2012) Tetrahedron 68(35):6999–7018
4. Riley DL, van Otterlo WAL (2011) Oxepins and azepines. In: Majumdar KC, Chattopadhay SK (eds) Heterocycles in natural product synthesis. Wiley, Weinheim, pp 537–565
5. Nicolaou KC, Aversa RJ (2011) Isr J Chem 51(3–4):359–377
6. Nicolaou KC, Frederick MO, Aversa RJ (2008) Angew Chem Int Ed 47(38):7182–7225
7. Hewitt RJ, Harvey JE (2010) J Org Chem 75(3):955–958
8. Dey S, Jayaraman N (2012) Beilstein J Org Chem 8:522–527
9. Sugita Y, Kimura C, Hosoya H, Yamadoi S, Yokoe I (2001) Tetrahedron Lett 42 (6):1095–1098
10. Batchelor R, Harvey JE, Northcote PT, Teesdale-Spittle P, Hoberg JO (2009) J Org Chem 74 (20):7627–7632
11. Furuta H, Hasegawa Y, Mori Y (2009) Org Lett 11(19):4382–4385
12. Furuta H, Hasegawa Y, Hase M, Mori Y (2010) Chem Eur J 16(25):7586–7595
13. Inagaki T, Nakamura Y, Sawaguchi M, Yoneda N, Ayuba S, Hara S (2003) Tetrahedron Lett 44(21):4117–4119
14. Abo T, Sawaguchi M, Senboku H, Hara S (2005) Molecules 10:183–189
15. Raju BC, Saidachary G, Saidachary G, Kumar JA (2012) Tetrahedron 68(31):6289–6297
16. Roux L, Charrier C, Salomon E, Ilhan M, Bisseret P, Tarnus C (2011) Tetrahedron Lett 52 (20):2586–2889
17. Pavlik C, Onorato A, Castro S, Morton M, Peczuh M, Smith MB (2009) Org Lett 11 (16):3722–3725
18. Raghavan S, Babu VS (2011) Chem Eur J 17(30):8487–8494
19. Yus M, Soler T, Foubelo F (2002) Tetrahedron 58(35):7009–7016
20. Lillo VJ, Gomez C, Yus M (2009) Tetrahedron Lett 50(19):2266–2269
21. Singh SA, Kabiraj S, Khandare RP, Nalawade SP, Upar KB, Bhat SV (2009) Synth Commun 40(1):74–80
22. Brunel Y, Rousseau G (1996) J Org Chem 61(17):5793–5800
23. Codelli JA, Puchlopek ALA, Reisman SE (2012) J Am Chem Soc 134(4):1930–1933
24. Alcazar E, Pletcher JM, McDonald FE (2004) Org Lett 6(21):3877–3880
25. Koo B, McDonald FE (2007) Org Lett 9(9):1737–1740
26. Matsumura R, Suzuki T, Sato K, Oku K-i, Hagiwara H, Hoshi T, Ando M, Kamat VP (2000) Tetrahedron Lett 41(40):7701–7704
27. Matsumura R, Suzuki T, Hagiwara H, Hoshi T, Ando M (2001) Tetrahedron Lett 42 (8):1543–1546
28. McDonald FE, Wang X, Do B, Hardcastle KI (2000) Org Lett 2(18):2917–2919
29. McDonald FE, Bravo F, Wang X, Wei X, Toganoh M, Rodriguez JR, Do B, Neiwert WA, Hardcastle KI (2002) J Org Chem 67(8):2515–2523
30. Bravo F, McDonald FE, Neiwert WA, Hardcastle KI (2004) Org Lett 6(24):4487–4489
31. Piccialli V, Borbone N, Oliviero G (2007) Tetrahedron Lett 48(29):5131–5135
32. Nicolaou KC, Hwang C-K, Duggan ME, Nugiel DA, Abe Y, Reddy KB, DeFrees SA, Reddy DR, Awartani RA, Conley SR, Rutjes FPJT, Theodorakis EA (1995) J Am Chem Soc 117 (41):10227–10238
33. Tripathi D, Kumar P (2008) Tetrahedron Lett 49(49):7012–7014
34. Tripathi D, Pandey SK, Kumar P (2009) Tetrahedron 65(11):2226–2231
35. Carreño MM, Des Mazery R, Urbano A, Colobert F, Solladié G (2004) Org Lett 6 (2):297–299
36. Lin J, Zhang W, Jiang N, Niu Z, Bao K, Zhang L, Liu D, Pan C, Yao X (2008) J Nat Prod 71 (11):1938–1941
37. Wang Y, Chen Y, He Q, Xie Y, Yang C (2013) Helv Chim Acta 96(2):296–308
38. Choi YL, Lim HS, Lim HJ, Heo J-N (2012) Org Lett 14(19):5102–5105
39. Sun LL, Hu BL, Tang R-Y, Deng C-L, Zhang X-G (2013) Adv Synth Catal 355 (2–3):377–382
40. Canoa P, Pérez M, Covela B, Gomez G, Fall Y (2007) Tetrahedron Lett 48(19):3441–3443

41. Garcia I, Pérez M, Gomez G, Fall Y (2008) Tetrahedron Lett 49(8):1344–1347
42. Grant TN, Benson CL, West FG (2008) Org Lett 10(18):3985–3988
43. Paduraru MP, Wilson PD (2003) Org Lett 5(25):4911–4913
44. Reddy CR, Ramesh P, Rao NN, Ali SA (2011) Eur J Org Chem (11): 2133–2141
45. Purino MA, Ramirez MA, Daranas AH, Martin VS, Padron JI (2012) Org Lett 14 (23):5904–5907
46. Yadav JS, Reddy BVS, Yadav NN, Satheesh G (2008) Synthesis (17): 2733–2738
47. Castaldi MP, Troast DM, Porco JA Jr (2009) Org Lett 11(15):3362–3365
48. O'Neil KE, Kingree SV, Minbiole KPC (2005) Org Lett 7(3):515–517
49. Crisostomo FRP, Martin T, Martin VS (2004) Org Lett 6(4):565–568
50. Yamamoto Y, Yamada J-I, Kadota I (1991) Tetrahedron Lett 32(48):7069–7072
51. Kuranaga T, Ohtani N, Tsutsumi R, Baden DG, Wright JLC, Satake M, Tachibana K (2011) Org Lett 13(4):696–699
52. Kadota I, Abe T, Ishitsuka Y, Touchy AS, Nagata R, Yamamoto Y (2007) Tetrahedron Lett 48(2):219–221
53. Taillier C, Gille B, Bellosta V, Cossy J (2005) J Org Chem 70(6):2097–2108
54. Fleck M, Bach T (2010) Chem Eur J 16(20):6015–6032
55. Lemaire P, Balme G, Desbordes P, Vors J-P (2003) Org Biomol Chem 1(23):4209–4219
56. Martin DBC, Vanderwal CD (2011) Chem Sci 2(4):649–651
57. Jepsen TH, Larsen M, Jorgensen M, Bronsted Nielsen M (2012) Synlett 23(3):418–422
58. Nishimura K, Kinugawa M (2012) Org Process Res Dev 16(2):225–231
59. Bosch J, Bachs J, Gomez AM, Griera R, Ecija M, Amat M (2012) J Org Chem 77 (14):6340–6344
60. Jones SB, Simmons B, Mastracchio A, MacMillan DWC (2011) Nature 475(7355):183–188
61. Lautens M, Paquin JF, Piguel S (2002) J Org Chem 67(11):3972–3974
62. Yoshida H, Asatsu Y, Mimura Y, Ito Y, Ohshita J, Takaki K (2011) Angew Chem Int Ed 50 (41):9676–9679
63. Parthasarathy K, Han H, Prakash C, Cheng C-H (2012) Chem Commun 48(52):6580–6582
64. Alcaide B, Almendros P, Martinez del Campo T (2007) Angew Chem Int Ed 46 (35):6684–6687
65. Rainier JD (2010) Synthesis of natural products containing medium-size oxygen heterocycles by ring-closing alkene metathesis. In: Cossy J, Arseniyadis S, Meyer C (eds) Metathesis in natural product synthesis: strategies, substrates and catalysts. Wiley-VCH, Weinheim
66. Crimmins MT, DeBaillie AC (2003) Org Lett 5(17):3009–3011
67. Bruder M, Haseler PL, Muscarella M, Lewis W, Moody CJ (2010) J Org Chem 75 (2):353–358
68. Chang M-Y, Wu M-H, Lee T-W (2012) Tetrahedron 68(31):6224–6230
69. Chang M-Y, Lee N-C (2012) Synlett 23(6):867–872
70. Doveston RG, Steendam R, Jones S, Taylor RJK (2012) Org Lett 14(4):1122–1125
71. Gross U, Nieger M, Bräse S (2010) Chem Eur J 16(38):11624–11631
72. He H, Ye K-Y, Wu Q-F, Dai L-X, You S-L (2012) Adv Synth Catal 354(6):1084–1094
73. Schmidt B, Biernat A (2008) Org Lett 10(1):105–108
74. Basu S, Ellinger B, Rizzo S, Deraeve C, Schürmann M, Preut H, Arndt H-D, Waldmann H (2011) Proc Natl Acad Sci USA 108(17):6805–6810
75. Shing TKM, Zhong Y-L (2006) Synlett (8): 1205–1208
76. Chatterjee A, Bhattacharya PK (2006) J Org Chem 71(1):345–348
77. Hofmann B, Reissig H-U (1993) Synlett (1): 27–29
78. Leyhane AJ, Snapper ML (2006) Org Lett 8(23):5183–5186
79. Sasaki M, Noguchi T, Tachibana K (1999) Tetrahedron Lett 40(7):1337–1340
80. Lee S-H, Van HTM, Yang SH, Lee K-T, Kwon Y, Cho W-J (2009) Bioorg Med Chem Lett 19 (9):2444–2447
81. Cong Z, Miki T, Urakawa O, Nishino H (2009) J Org Chem 74(10):3978–3981
82. Corser DA, Marples BA, Dart RK (1992) Synlett (12): 987–989

83. Kas'yan LI, Tarabara IN, Kas'yan AO (2004) Russ J Org Chem 40(9):1227–1257
84. Matsuo G, Hori N, Nakata T (1999) Tetrahedron Lett 40(50):8859–8862
85. Nakata T (2010) Chem Rec 10(3):159–172
86. Bamoharram FF, Heravi MM, Roshani M, Gharib A, Jahangir M (2006) J Mol Catal A Chem 252(1–2):90–95
87. Endo Y, Bäckvall J-E (2011) Chem Eur J 17(45):12596–12601
88. Ito M, Osaku A, Shiibashi A, Ikariya T (2007) Org Lett 9(9):1821–1824
89. Ebine M, Suga Y, Fuwa H, Sasaki M (2010) Org Biomol Chem 8(1):39–42
90. Nicolaou KC, Aversa RJ, Jin J, Rivas F (2010) J Am Chem Soc 132(19):6855–6861
91. Marek A, Patil MR, Klepetarova B, Kohout L, Elbert T (2012) Tetrahedron Lett 53 (16):2048–2050
92. Ghosh AK, Xi K (2009) J Org Chem 74(3):1163–1170
93. Snyder SA, Treitler DS, Brucks AP (2011) Aldrichimica Acta 44(2):27–40
94. Snyder SA, Wright NE, Pflueger JJ, Breazzano SP (2011) Angew Chem Int Ed 50 (37):8629–8633
95. Parenty A, Moreau X, Campagne J-M (2006) Chem Rev 106(3):911–939
96. Basabe P, Bodero O, Marcos IS, Diez D, Blanco A, de Roman M, Urones JG (2009) J Org Chem 74(20):7750–7754
97. Fuwa H, Noji S, Sasaki M (2010) J Org Chem 75(15):5072–5082
98. Altemöller M, Gehring T, Cudaj J, Podlech J, Goesmann H, Feldmann C, Rothenberger A (2009) Eur J Org Chem (11): 2130–2140
99. Cudaj J, Podlech J (2012) Synlett 23(3):371–374
100. Zhao W, Li Z, Sun J (2013) J Am Chem Soc 135(12):4680–4683
101. Wang L, Thai K, Gravel M (2009) Org Lett 11(4):891–893
102. Fujiwara H, Kurogi T, Okaya S, Okano K, Tokuyama H (2012) Angew Chem Int Ed 51 (52):13062–13065
103. Ebine M, Fuwa H, Sasaki M (2011) Chem Eur J 17(49):13754–13761

Synthesis of Eight- to Ten-Membered Ring Ethers

Elena Soriano and José Marco-Contelles

Contents

Abstract Polyfunctionalized medium-sized cyclic ethers have attracted much attention from synthetic and medicinal chemists due to their presence in a wide range of biologically active natural products. Accordingly, much attention has been focused on efficient approaches towards these systems. However, their synthesis is generally difficult via standard cyclization methods due to the ring size. In this chapter, we summarize the most important and successful approaches and strategies to prepare eight- to ten-membered oxacycles, including oxocanes, oxonanes, and oxecanes, and their applications to the synthesis of natural products containing these structural motifs.

E. Soriano (✉)
Síntesis, Estructura y Propiedades de Compuestos Orgánicos (SEPCO), Instituto de Química Orgánica General (CSIC), Juan de la Cierva, 3, 28006 Madrid, Spain
e-mail: esoriano@iqog.csic.es

J. Marco-Contelles
Laboratorio de Química Médica, Instituto de Química Orgánica General (CSIC), Juan de la Cierva, 3, 28006 Madrid, Spain
e-mail: iqoc21@iqog.csic.es

J. Cossy (ed.), *Synthesis of Saturated Oxygenated Heterocycles II*, Topics in Heterocyclic Chemistry 36, DOI 10.1007/978-3-642-41470-1_2,

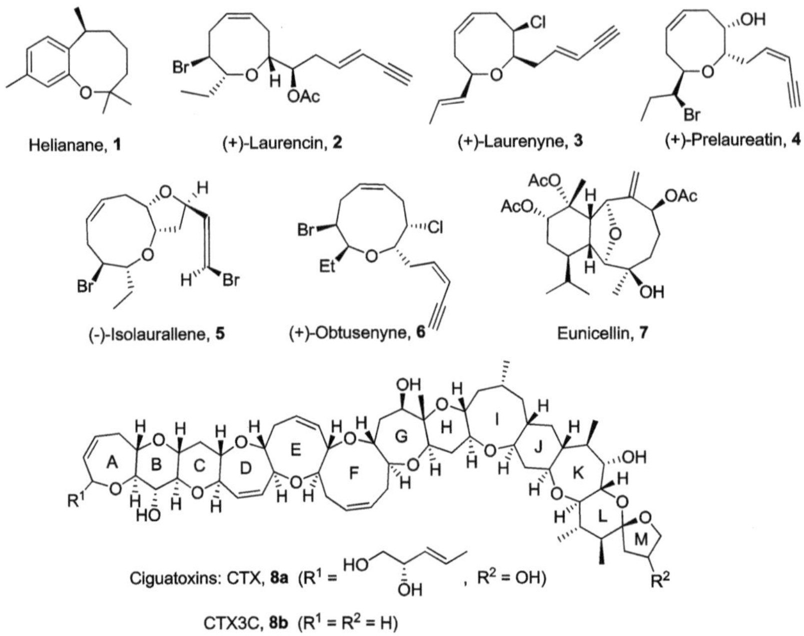

Fig. 1 Representative examples of medium-sized ethers

Keywords Heterocyclization • Medium-ring ethers • Natural products • Oxacycles • Synthesis

1 Introduction

Medium-sized cyclic ethers have attracted much attention from synthetic and medicinal chemists due to their presence in a wide range of biologically active natural products, including marine ladder toxins (brevetoxins, ciguatoxins, etc.), lauroxanes, and heliannuol A, along with members of the eunicellin family, or antibiotics [1]. Examples of their occurrence in Nature include isolaurepinnacin, laurencin, obtusenyne, brevetoxin, and many others (Fig. 1). In view of the increasing number of biologically active natural products containing medium- and large-sized cyclic ether derivatives, much attention has been focused on efficient approaches towards these systems. However, the preparation of these skeletons is generally difficult by standard cyclization methods [2] and remains a significant challenge, primarily because both entropic (probability of the chain ends meeting) and enthalpic (increasing strain in the transition state) barriers hamper cyclization strategies [3]. Nevertheless, the challenge in their efficient construction has led to

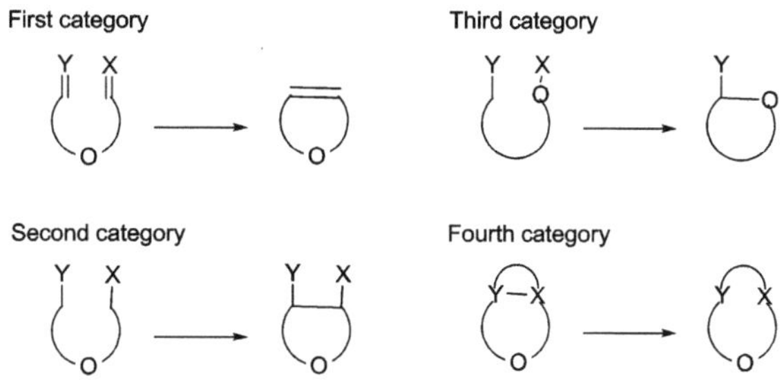

Fig. 2 Categories for medium-sized ether formation

the development of several strategies for their synthesis in recent years [4, 5]. Several excellent reviews about the medium-sized ether syntheses have been published already [6–16]. In this chapter, the syntheses of 8-oxocine, 9-oxonene, and 10-oxocine-membered rings will be reviewed.

2 Methods and Applications

According to Hoberg's classification [17], the construction of medium-sized ethers can be classified into the following basic categories (Fig. 2). The first category is related to the carbon–carbon double bond formation, and the focus will be on the olefination through the ring-closing metathesis (RCM), which provides highly efficient ways for medium-sized ether formation. The second category involves a carbon–carbon single bond formation. The third category involves the formation of carbon–oxygen single bond. The fourth category involves ring expansion of different ring size as well as rearrangements. As a closely related method, the epoxide rearrangements and ring openings can be additionally discussed. Although the conversion of lactones into ethers can be considered, it will not be included in this review since the discussion on the lactonization methods is beyond the scope of this chapter and has recently been reviewed and updated [18, 19].

In this chapter, special attention will be paid to the RCM method since this has been widely used in the total syntheses of a plethora of natural products. Accordingly, the pertinent section has been divided in three subsections to describe separately the application to the synthesis of eight, nine, and ten membered-ring oxacycles.

2.1 Metathesis to Form Carbon–Carbon Double Bond

The discovery and development of olefin metathesis over the past few years have revolutionized the area of medium-ring synthesis. Thus, ring-closing metathesis (RCM) of dienes is one of the most important reactions to construct cyclic

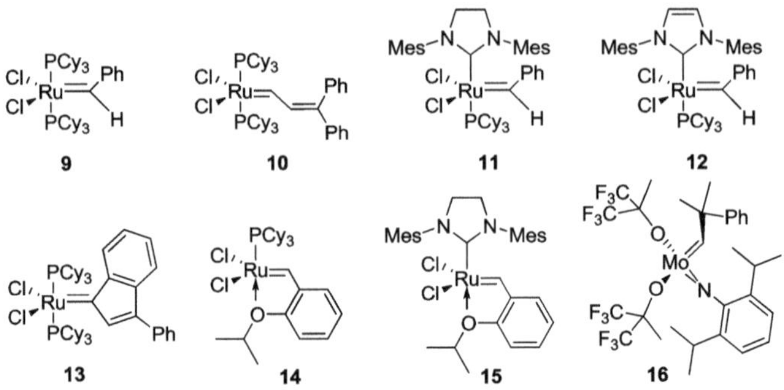

Fig. 3 Examples of metathesis catalysts

compounds and has frequently been utilized to make medium-sized ethers [20, 21]. Several reviews have been published in this area in recent years [22–27].

The most common catalysts for this reaction include the Schrock's molybdenum complex, Grubbs ruthenium complexes (first and second generation), and the Hoveyda–Grubbs ruthenium complex (Fig. 3). The Schrock catalyst is highly reactive but has a low functional group tolerance and stability. The Grubbs first-generation catalyst is less reactive but has a high functional group tolerance and stability. The Grubbs second-generation catalyst is very reactive and thermally stable. Likewise, the Hoveyda–Grubbs catalyst is very reactive and stable. Dichloromethane, dichloroethane, benzene, and toluene are frequently used as solvents. The reaction temperatures vary from room temperature to 110 °C. Such reaction conditions are quite mild and compatible with many sensitive functional groups including esters, amides, ketones, ethers, acetals, etc. The ring sizes that can be formed from the RCM are mainly limited to five-, six-, and seven-membered rings and large macrocycles, while three-, four-, and eight- to eleven-membered rings are always difficult to construct [28]. Recent publications concerning the formation of medium-sized ethers via RCM have been reported [29–39]. In most cases, RCM can be applied to ω-dienes to construct the medium-sized ethers. However, when alkenes were either sterically hindered or electronically deactivated, relay ring-closing metathesis (RRCM) was utilized [40–42].

The formation of medium-sized oxacycles by RCM constitutes a considerable challenge, since the inherent ring strain predisposes cycloalkenes containing 8–11 atoms towards ring-opening metathesis or ring-opening metathesis and polymerization. However, the RCM was successful for the synthesis of a variety of these cycloalkenes. In order to facilitate the desired cyclization, several features can be installed in the substrate providing some sort of conformational constraint. The presence of these beneficial groups, however, does not always guarantee the success of the cyclization. It has been reported that these constraints can be induced by using either preexisting rings (cyclic conformational constraints) or acyclic conformational constraints. Thus, for instance, if the olefinic side chains are positioned on the ring vicinal to each other, the cyclization is very easy. Examples include the *trans*-fused polyether systems **17** (Scheme 1) [43, 44]. Only for the nine-membered ring,

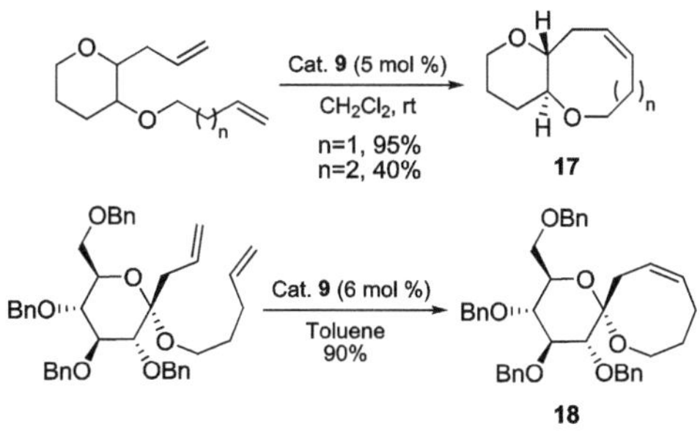

Scheme 1 Grubbs catalyst **9** for the preparation of oxacyclene

a decrease of the yield is coming from steric hindrance in the transition state. In addition, the oxygen atoms seem not to interfere with the catalytic cycle. Eight-membered ring could be additionally attached to a sugar derivative in a spirocyclic fashion as in **18** [45].

In order to limit the conformational freedom, appropriate positioning of the substituents on the chain possessing the two olefinic groups can boost the cyclization. An oxygen-bearing vicinal stereocenter implies a suitable conformation of the molecule for RCM. This conformation is also known as a *gauche* conformation. Thus, Crimmins et al. reported the *gauche effect* of 1,2-dioxygen substituents to facilitate the ring closure (Scheme 2). They proved the power of the RCM approach by cyclizing chiral ethers. For instance, dienes **20**, prepared by the Evans aldolization of glycolate imides with acrolein, followed by reduction and acetylation, were cyclized to **21** with high yield by using the Grubbs catalyst **9**. It was suggested that the vicinal stereocenters in intermediate **20** induce a conformation where the olefinic chains are positioned in a gauche conformation [46, 47].

2.1.1 Synthesis of Eight-Membered Oxacycles

The concept of the *gauche* effect has been applied to the synthesis of a key precursor of (+)-laurencin by Crimmins and Choy [46]. When compounds **22** and **24** were exposed to [Ru]-catalyst **9**, the corresponding RCM products **23** and **25** were obtained in good yields. The facile formation of the RCM product is due to the two synergistic gauche effects by a pair of vicinal substituents on the dienes (Scheme 3).

The marine red algae genus *Laurencia* specifically produces medium-ring ethers as secondary metabolites, which typically possess a linear C15 carbon skeleton, one or several halogen atoms adjacent to an ethereal oxygen atom, and an enyne or a bromoallene side chain [48]. Since the first report of (+)-laurencin, **2**, isolated from *Laurencia glandulifera* by Irie et al. [3], numerous medium-ring ethers have been isolated from *Laurencia* red algae. These *Laurencia* oxacycles have attracted the

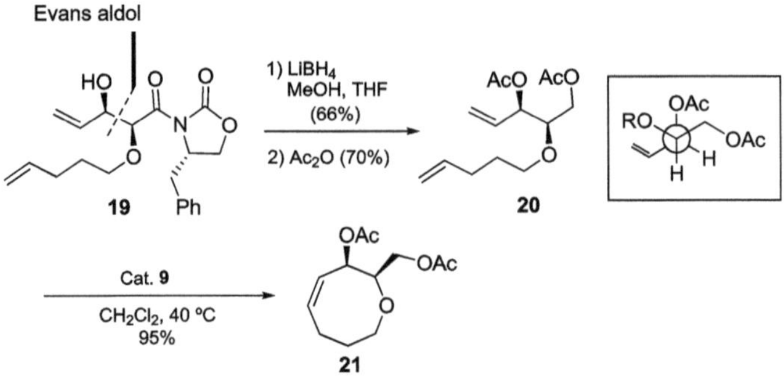

Scheme 2 Use of the gauche effect of 1,2-dioxygen substituents to enhance the ring-closing metathesis

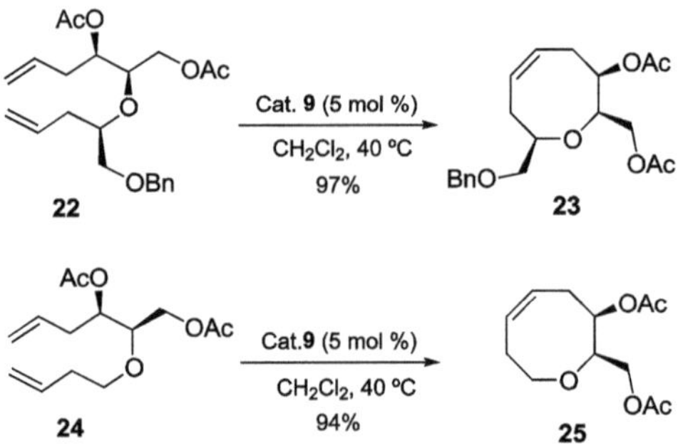

Scheme 3 Application of the gauche effect to the synthesis of a key precursor of (+)-laurencin

attention of synthetic chemists for a long time because their unusual structural features have provided synthetic challenges. Since the pioneering first total synthesis of **2**, achieved by Masamune et al. [49], other *Laurencia* natural medium-ring ethers have been synthesized. So far, diverse contributions to the total and formal total syntheses have been reported [50–59].

Crimmins and Emmitte [60] have also published a direct synthesis of (+)-laurencin, in which the side chain was included early in the synthesis (Scheme 4). Instead of aldolizations, asymmetric alkylations were used to set up two of the stereocenters. Thus, diol **26** was prepared by allylation of a benzyl glycolate imide followed by a chelation-controlled ethynylation of the derived aldehyde. Another allylation of **27** provided diene **28** that cyclized to oxocene **29**. Further steps were needed to change the protecting group, generate the aldehyde **30**, and homologate the side chain. The addition of the chlorotitanium enolate of (*S*)-(+)-3-acetyl-4-isobutyl-2-thiazolidininone gave a diastereomeric mixture of alcohols **31** in a ratio of 3:3:1. Finally, a Wittig reaction led to the natural product (+)-laurencin, **2**.

Further reports from Crimmins et al. described the first total synthesis of (±)-prelaureatin, **4**, 3*E*-prelaureatin, and (+)-laurallene, demonstrating that both the laurenan and lauthisan derivatives are accessible through similar asymmetric glycolate aldol ring-closing metathesis sequence, exploiting the *gauche* effect [61]

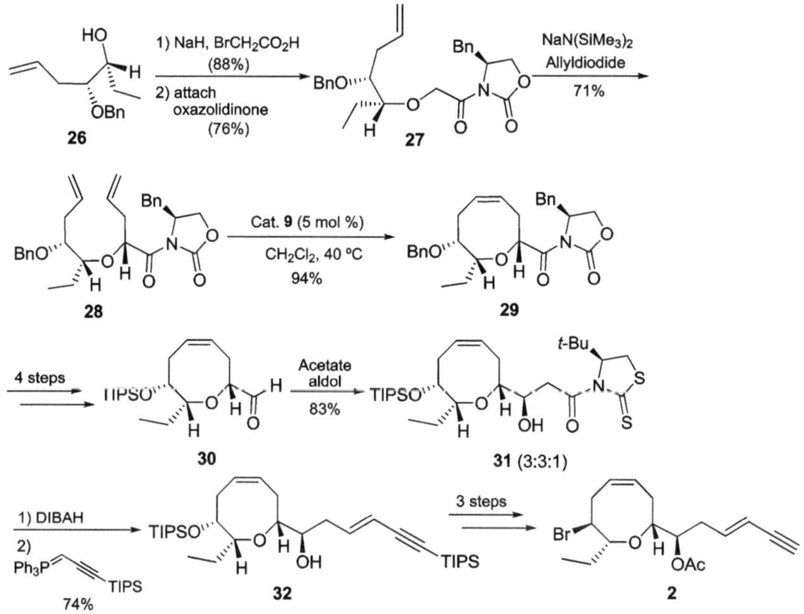

Scheme 4 Crimmins et al. total synthesis of (+)-laurencin, **2**

of the C6 and C7 oxygen substituents to accelerate the ring closure, thus illustrating the efficiency and power of this protocol for the construction of complex medium-ring ethers [57].

Kim et al. have achieved a concise total synthesis of (+)-laurencin, **2**, based on an intermolecular allylation RCM strategy [62, 63]. Fujiwara–Murai et al. have accomplished the total synthesis of **2** from a sugar derivative (β-D-galactose pentaacetate) [56], showing that the combined use of ether synthesis and ring-closing olefin metathesis with second-generation Grubbs catalyst provided an efficient synthetic strategy for medium-ring ethers [34, 64]. In 2002, this group also reported a successful application of this combination of RCM and *C*-glycoside ring cleavage in the total synthesis of (+)-prelaureatin, **4** [65]. The total synthesis, shown in Scheme 5, started from β-D-galactose pentaacetate which was transformed to **33**. Protection of **33** as a benzyl ether followed by desilylation and Swern oxidation gave **34** [47], which was allylated according to Grieco procedure [66] providing **35** stereoselectively. The *C*-glycoside **35** was subjected to a ring-cleavage process including benzyl protection, removal of the acetonide group, oxidative cleavage of the diol moiety, and reduction of the resulting dialdehyde to produce acyclic diol **36**. After mesylation of **36**, followed by removal of the benzyl groups with DDQ [67], the resulting dimesylate ester was transformed to **37** through a basic treatment followed by TBS protection. Selective methylation of the epoxide of **37** and subsequent cyanation of the mesylate ester provided **38**, which was converted to **39** by bromination and formation of dibromoolefin. Although the RCM of **39** with the second-generation Grubbs catalyst mainly gave cyclohexene **41**, the cyclization with the first-generation Grubbs catalyst

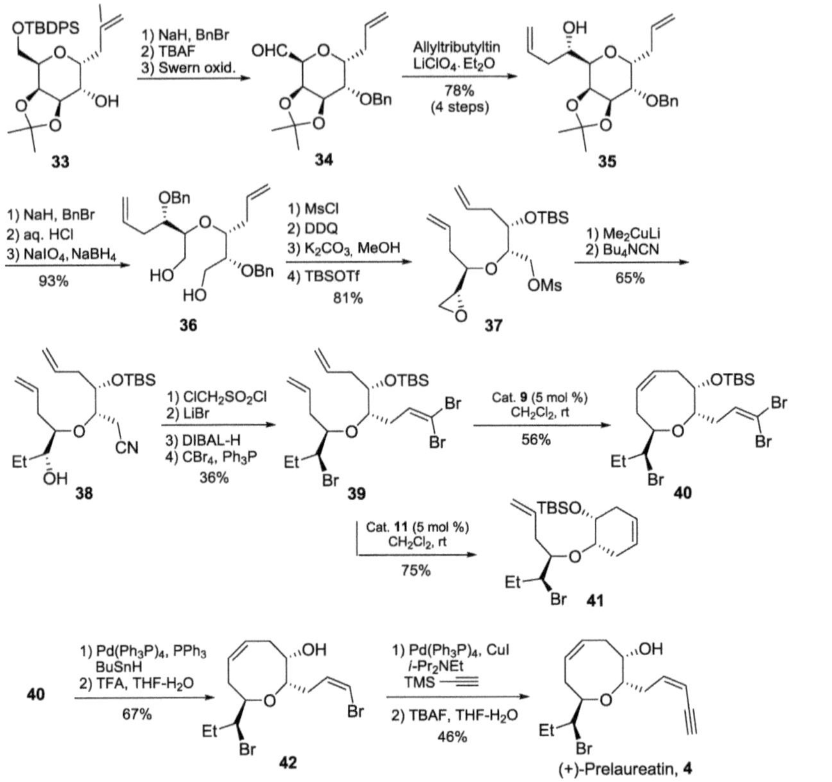

Scheme 5 Fujiwara–Murai et al. total synthesis of (+)-prelaureatin, **4**

produced oxocene **40** as a major product. Partial hydrogenolysis of dibromoolefin **40** in accordance with the Uenishi procedure [68], followed by desilylation, afforded **42**, which was subjected to a Sonogashira reaction and desilylation to furnish (+)-prelaureatin, **4**.

Taylor et al. [47] confirmed the remarkable effect of the protecting group in the formation of eight-membered rings by RCM. The ease to form eight-membered heterocycle **44** was favored as the bulkiness of the protecting group increased. Without hydroxyl protection, substrate **43** gave no RCM product. Surprisingly, substrate **45** containing the bulky TBS-protecting groups furnished the RCM product **46** in 90 % yield (Scheme 6).

In some cases, conformational tuning of the RCM precursor is achieved by incorporating a preexisting ring, such as a benzene ring or a carbo- or heterocyclic ring [69]. The presence of an additional ring can act as a conformational constraint, which aids the conformational reorganization of the diolefinic substrate and thereby facilitates the RCM. In addition, the oxygen atom present in cyclic ethers is known to induce conformational constraints in the molecule. An example of these observations was reported by Clark and Kettle within their synthetic studies towards

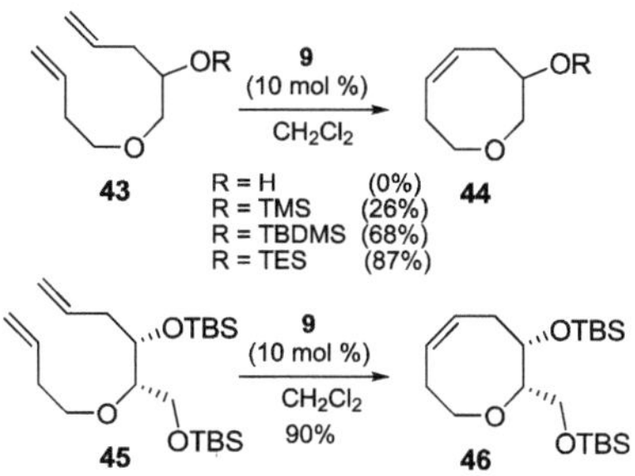

Scheme 6 Effect of protecting groups on the RCM

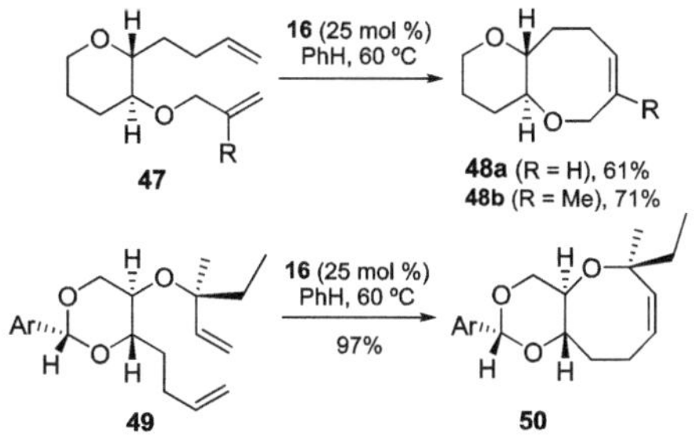

Scheme 7 Conformational constraint in RCM by additional rings

brevetoxin A. Compounds **47** and **49** can be easily converted into the corresponding cyclized products **48** and **50** under the influence of the Schrock catalyst **16** (Scheme 7) [70]. RCM, in the formation of oxo-ring systems, were hampered by isomerization and/or dimerization [71]. In this situation, the addition of the catalyst in several small portions proved to be beneficial [72]. At the same time, Clark et al. also pointed out that the cyclization via ring-closing enyne metathesis (RCEM), rather than RCM of ω-dienes, was sluggish and low yielding [73].

The relative stereochemistry of the 1,2-substituents on a six-membered ring has profound influence on the rate of the RCM reaction. Grubbs et al. [74] demonstrated that the *trans*-stereochemistry facilitates the synthesis of the [6.4.0] system. Thus, *trans*-diene **51a** was transformed in 60–75 % to the cyclized product, whereas for the corresponding *cis*-diene **51b**, only 20–33 % of the cyclic ether was obtained (Scheme 8).

It was reported that the tributylstannyl group could be used as a large group to affect a favorable conformation for the synthesis of an oxocine ring by RCM from an acyclic diene precursor [75]. RCM of dienes **53** (Scheme 9) with catalyst **10** led to the cyclic ethers **54** in good to excellent yields. The α-(alkoxyalkyl)stannane moiety can further undergo transmetallation by lithio-destannylation, and the intermediate carbanion could be trapped by electrophiles to provide the substituted oxocines. When the tributylstannyl group is replaced by a *tert*-butyl group, the diene undergoes polymerization under similar conditions. This has been attributed

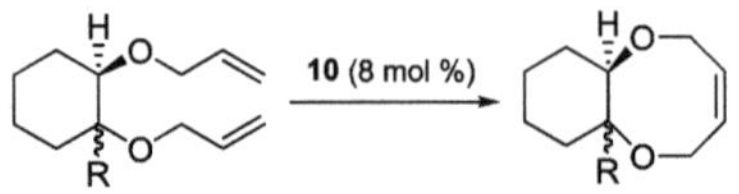

Scheme 8 Effect of the relative stereochemistry of the substituents

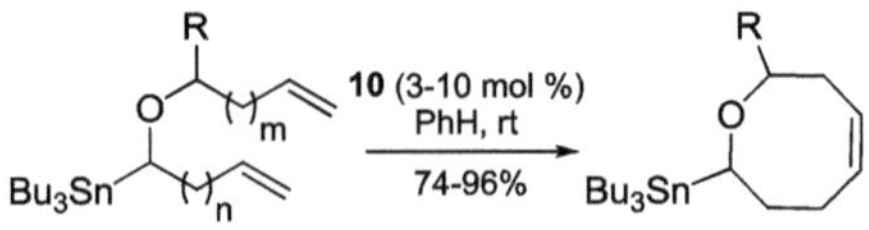

Scheme 9 Effect of tributylstannyl group on the conformation in RCM

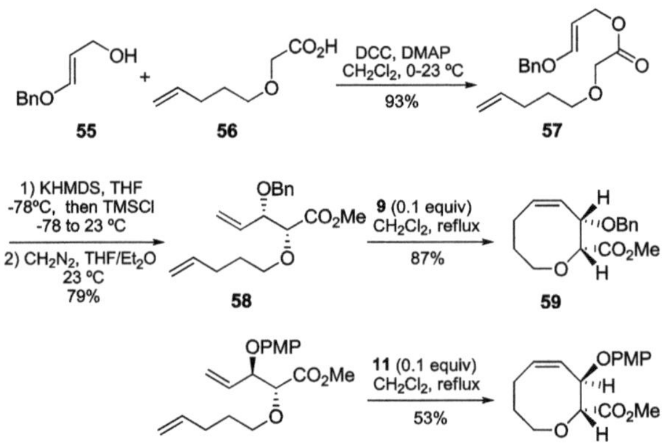

Scheme 10 Claisen–Ireland rearrangement and RCM

to the larger size of the tributylstannyl group compared to the *tert*-butyl group. Another reason for the enhancement of the reactivity of the diene can also be a stereoelectronic effect due to the interaction between Sn and O atoms or the Sn atom and the Ru carbene complex.

A simple stereoselective synthesis of *cis*- and *trans*-2,3-disubstituted medium-sized cyclic ethers **59** and **61** (Scheme 10) has been developed [30] based on the Ireland–Claisen rearrangement and RCM. The glycolate ester **57** was prepared by a condensation of glycolic acid **56** with (*E*)-3-benzyloxy-2-propenol **55**. Deprotonation of **57** with KHMDS at −78 °C for 5 min followed by treatment with TMSCl and warming to rt induced an Ireland–Claisen rearrangement and, after esterification with diazomethane ester **58**, was isolated in 79 % overall yield. The RCM using the first-generation Grubbs catalyst **9** provided the eight-membered

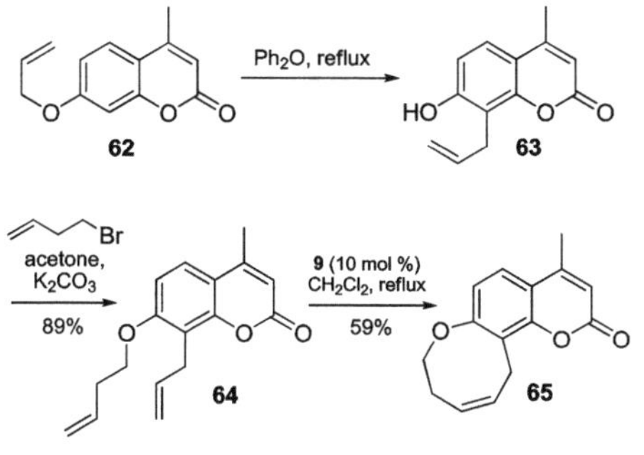

Scheme 11 Combination of the Claisen rearrangement with RCM for the preparation of medium-sized ethers

cyclic ether **59** in 87 % yield. Similarly, diene **60** smoothly underwent a RCM with the second-generation Grubbs catalyst **11** to give the cyclic ether **61** in 53 % yield.

Chattopadhyay et al. combined [76, 77] a Claisen rearrangement with a RCM for the synthesis of a range of carbocycles and heterocycles. Thus, 8-allyl-7-hydroxy-4-methylcoumarin (**63**), obtained by using a Claisen rearrangement of 7-allyloxy-4-methylcoumarin (**62**) (Scheme 11), was alkylated with butenyl bromide to install the diene, producing **64**. When compound **64** was treated with Grubbs catalyst **9**, the oxocinocoumarin **65** was obtained. Similarly, several other oxepino- and oxocinocoumarins were analogously prepared. The study was later extended to the preparation of oxepine and oxocine-annulated 2-quinolones [78].

The heliannuols are a promising group of phenolic allelochemicals isolated from *Helianthus annuus* that exhibit useful biological activity [79]. The benzooxocin ring system present within these molecules has induced synthetic activity using RCM for their preparation. Snieckus and Stefinovic first reported [80] that the diallylated benzene derivative **66** (Scheme 12), prepared by using their directed-metallation protocol, when treated with catalyst **9** followed by a catalytic hydrogenation of **67** provided (±)-helianane. Shishido et al. reported [81] an enantioselective total synthesis of (−)-heliannuol A **73** (Scheme 13), the most active member of the family [82]. Diene **70** was prepared in a sequence of ten steps in which the stereogenic center at C7 was set by an enzymatic desymmetrization of the prochiral diol **68**. Compound **69** was obtained in 78 % enantiomeric excess, which was increased to 100 % by recrystallization. RCM of diene **70**, induced by catalyst **11**, provided oxocin derivative **71** in an impressive yield of 88 %. The high yield for this cyclization is probably due to the conformational constraints induced by both the benzene ring and the geminal dimethyl group [83]. Compound **71** was then elaborated into the natural product through a selective epoxidation.

At this point, and to highlight the potential of the metathesis strategy for the synthesis of medium-sized oxacycles, it should be mentioned that two structurally related families of natural products, although lactones, have been successfully synthesized by RCM. Recently, a sequential Evans–Tishchenko and RCM developed by Aird et al. [84] was used to access the functionalized eight-membered ring core of octalactins, highly functionalized compounds produced by an actinomycete collected

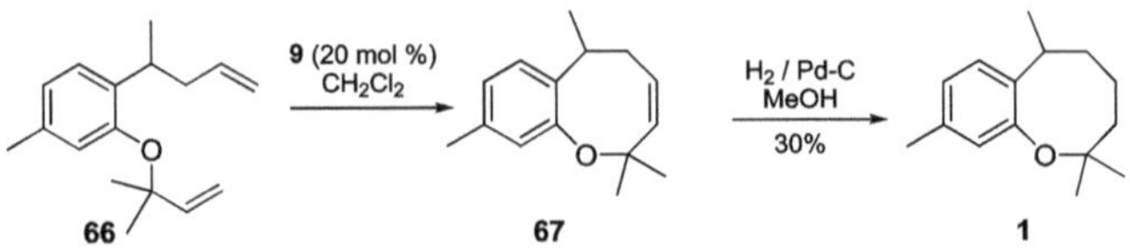

Scheme 12 Synthesis of the benzooxocin core of heliannuols

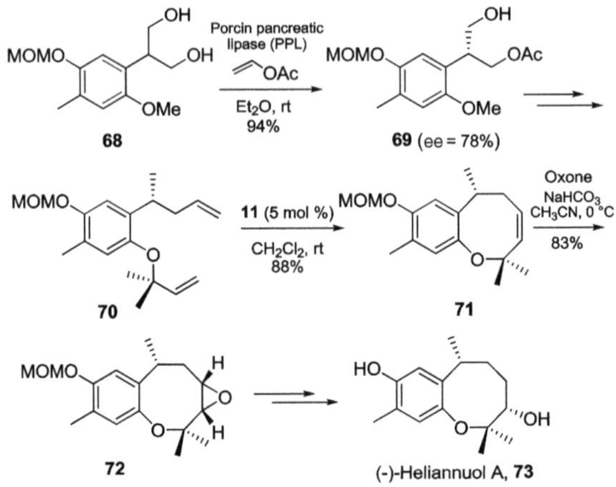

Scheme 13 Shishido et al. enantioselective total synthesis of (−)-heliannuol A, **73**

from the surface of a Mexican gorgonian octocoral of the genus *Pacifigorgia*. The synthesis initially relied on the preparation of chiral aldehydes **76** and **78** and phosphonate **82**, as outlined in Scheme 14. Horner–Wadsworth–Emmons reaction applied to **78** and **82**, followed by a desilylation, furnished the β-hydroxy enone **83** as a 95:5 mixture of the *E*/*Z*-isomers. Evans–Tishchenko coupling of fragment **83** with aldehyde **76**, using a preformed samarium(III) catalyst, led to dienoic ester **84** as a single diastereomer. A RCM applied to **84**, using the second-generation Grubbs catalyst and $Ti(O\textit{i}Pr)_4$ as Lewis acid, afforded a 1:1 mixture of the eight-membered lactone **86** and the cyclopentene derivative **85**. This latter compound was formed due to a competitive metathesis with the trisubstituted double bond and one of the monosubstituted double bond present in **84**.

The solandelactones A–I constitute a group of eight-membered ring heterocycles isolated in 1996 from *Solanderia secunda*, a dark-brown hydroid found in the shore of Jeju island, in Korea. Some members of this family of metabolites exhibit moderate inhibitory activity against FPT (farnesyl protein transferase) [85]. These compounds are lactonized cyclopropyl docosanoids, being probably the first examples of marine oxylipins bearing a carbon skeleton of 22 atoms. The original assignment of a skeleton of 11 carbon atoms of two solandelactones was recently revised by their total synthesis [86, 87]. An approach to the synthesis of the cyclopropyl lactone segment of solandelactones starting from

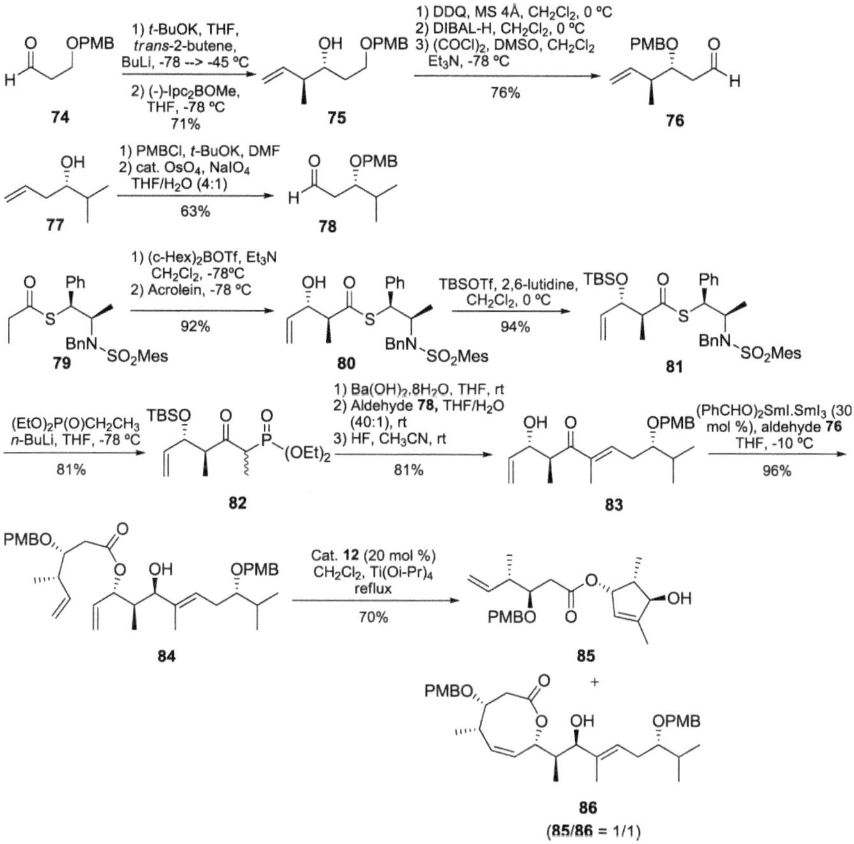

Scheme 14 Synthesis of the eight-membered ring core of octalactins developed by Aird et al.

2,3-*O*-isopropylidene-D-glyceraldehyde **87** and based on a RCM of the dienoic ester **91** was developed by Varadarajan et al. [88] (Scheme 15).

Finally, it should be noted that the RCM has been applied to the construction of eight-membered oxocin scaffold in polycyclic systems. Some examples are detailed below. In 1989, Yasumoto et al. determined the structure of ciguatoxin (CTX, **8a**, Fig. 1), a neurotoxic polycyclic ether produced by the marine dinoflagellate *Gambierdiscus toxicus* [89]. To date, the structures of more than 20 congeners including CTX3C **8b** [90] were also determined. The complex structure consists of five- to nine-membered polycyclic ether core (A-M rings) containing 30 chiral centers, a 5,6-spiroacetal, and four double bonds, three hydroxyl groups, and five methyl groups on the ether rings. The size and the structural and stereochemical complexity have made them irresistible synthetic targets, providing a stimulus for the development of many new reactions and strategies. The first total synthesis of CTX3C **8b** by Hirama et al. in 2001 has been achieved on the basis of a highly convergent strategy featuring the chemoselective RCM as one of the key steps

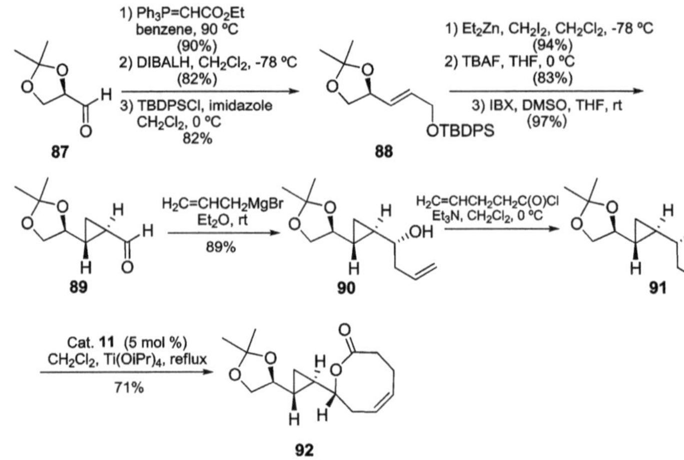

Scheme 15 Synthesis of the cyclopropyl lactone of solandelactone **92** developed by Varadarajan et al.

[91]. Later, they further improved the total synthesis and also reported a second-generation total synthesis of CTX3C in which the penultimate step involved a RCM-mediated construction of the central eight-membered O-ring [92]. The formation of the eight-membered E- and I-rings was carried out by RCM. Moreover, once the ABCDE-ring system and the HIJKLM-ring system were synthesized, the total synthesis was accomplished by the union of both segments by construction of the F-ring through a RCM.

Other groups have reported elegant total synthesis of CTX3C. Thus, Clark et al. have disclosed a relatively short and efficient approach to the synthesis of the pentacyclic ABCDE fragment of CTX3C in which two-directional and iterative RCM were used to effect the ring construction [93]. Kadota et al. achieved a convergent synthesis of the A–E-ring segment of ciguatoxin CTX3C via the intramolecular allylation of a chloroacetoxy ether and a RCM using the second-generation Grubbs catalyst to construct ring E [94]. Recently, this group has developed a convergent synthetic route to the HIJKLM-ring system of ciguatoxin CTX3C. The key transformations were a conjugate addition/alkylation, spiroacetalization, intramolecular allylation, and finally RCM, followed by a stereoselective hydrogenation, where the RCM was carried out with the Hoveyda–Grubbs second-generation catalyst to provide the corresponding cyclized I-ring [95]. Fujiwara et al. also described an efficient synthesis of the IJKLM-ring part of ciguatoxin CTX3C [34]. Their synthesis also featured a RCM as the key step, but at an earlier stage.

Brevetoxin A (Fig. 4) is a decacyclic ladder toxic metabolite of *Karenia brevis* possessing five-, six-, seven-, eight-, and nine-membered oxacycles, as well as 22 tetrahedral stereocenters. This compound is known to cause the infamous red tide phenomenon responsible for massive fish kills, as well as neurotoxic shellfish poisoning and bronchial irritation in humans. Crimmins et al. have developed a unified approach to the construction of the B-, E-, G-, and J-rings based upon a

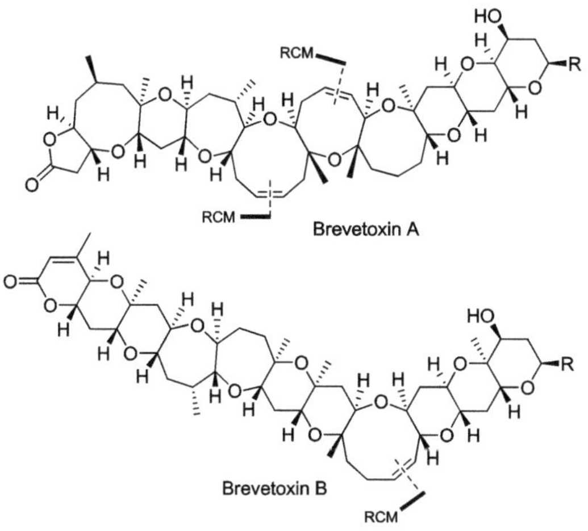

Fig. 4 Structure of brevetoxins A and B

RCM strategy from the corresponding diene at the end of a convergent synthesis of brevetoxin A [96]. Brevetoxin B, a potent neurotoxin, was isolated from the red tide organism *Gymnodinium breve* Davis in 1981 as the first example of marine polycyclic ethers. The total synthesis of brevetoxin B has been accomplished in a highly convergent manner by the assembly of three fragments, where the final cyclization to form the H-ring was realized by a RCM [97].

2.1.2 Synthesis of Nine-Membered Oxacycles

As noted previously, a sugar moiety induces conformational constraints that may favor the ring formation from a diene using a RCM. On the contrary, the helpful effect of geminal substitution for the formation of eight-membered rings (see Schemes 12 and 13) is negligible for the more flexible nine-membered rings. Thus, it has been reported that a sugar moiety present in **93** orients the two alkenyl moieties towards a suitable *gauche* conformation and thus facilitates the RCM reaction [98]. In this regard, the diene-containing precursor **93** was subjected to RCM and the corresponding nine-membered cycloalkene **94** was formed in a good yield (Scheme 16).

Martin and Delgado have developed a route to *trans*-fused bicyclic ethers from a disubstituted tetrahydropyran ring [99]. Further exploitation of the beneficial role of the *gauche* effect induced by vicinal dihydroxy groups, leading to the formation of nine-membered cyclic ethers, has been reported by Crimmins et al. during the synthesis of isolaurallene **5** (Scheme 17) [100]. Thus, diene **95** was subjected to cyclization with Grubbs catalyst **9** leading to oxacycle **96**. It was argued that diene **95** underwent such facile ring closure due to two synergistic gauche effects at C6–C7 and C12–C13. Compound **96** was then elaborated to (−)-isolaurallene **5**.

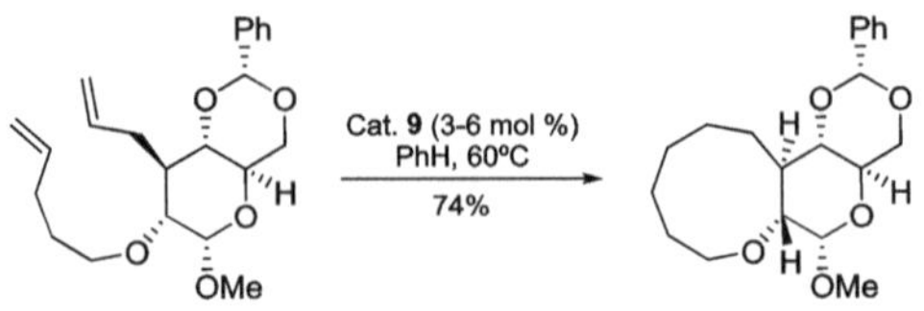

Scheme 16 Formation of nine-membered ring ethers showing the effect of a sugar moiety on the RCM

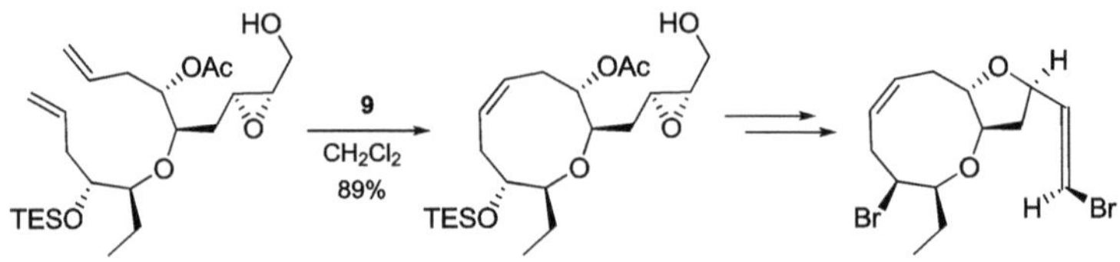

Scheme 17 Synthesis of isolaurallene **5** by Crimmins et al.

The same group also applied a similar RCM-based strategy for the enantioselective synthesis of (+)-obtusenyne, **6** [101], a member of a family of compounds isolated from *Laurencia* algae and sea hares. In this synthesis, the nine-membered ether was directly synthesized from an acyclic precursor by using a highly efficient RCM. The key steps were an asymmetric glycolate alkylation to establish the stereochemical relationship of the α,α'-disubstituted ether linkage and a subsequent RCM to construct the nine-membered oxocene (Scheme 18).

Kaliappan and Kumar reported [102], during the synthesis of the B-ring of eleutherobin (a potent cytotoxic compound), that the glucose-derived diene **105** (R = Me), prepared from **104** (Scheme 19), was transformed to the desired cyclized product **106** in 33 % when treated with catalyst **9**, the remainder being unreacted starting material and, presumably, cross-metathesis products. The observation of a similar result with the TBS ether suggested that the low yield could be attributed to the coordination of the metal center with the oxygen of the furanose ring. In order to destabilize this chelate structure, the RCM of dienes **105** was carried out with a catalytic amount of catalyst **9** in the presence of a substoichiometric amount of $Ti(O\textit{i}Pr)_4$. This modified protocol successfully led to the formation of the desired RCM products **106** in higher yields.

Following the approach developed by Chattopadhyay et al. described above, which combined [76, 77] a Claisen rearrangement with a RCM for the synthesis of a range of carbocycles and heterocycles, the synthesis of oxoninoquinolones has been carried out. Thus, Claisen rearrangement of the *O*-allyl ether **108** (Scheme 20), prepared from the hydroxyquinolone **107**, led to the rearranged phenol **109** which after an alkylation step led to dienes **110**, which were transformed to the oxoninoquinolones **111** in a very good yield after a RCM [103].

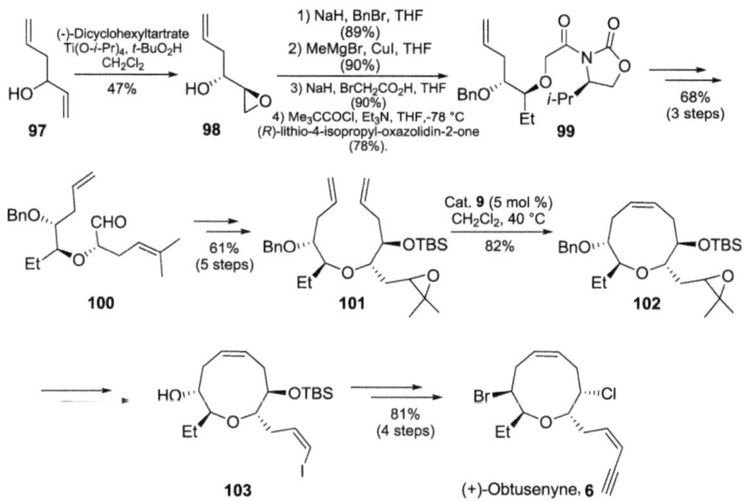

Scheme 18 Synthesis of (+)-obtusenyne, **6** by Crimmins et al.

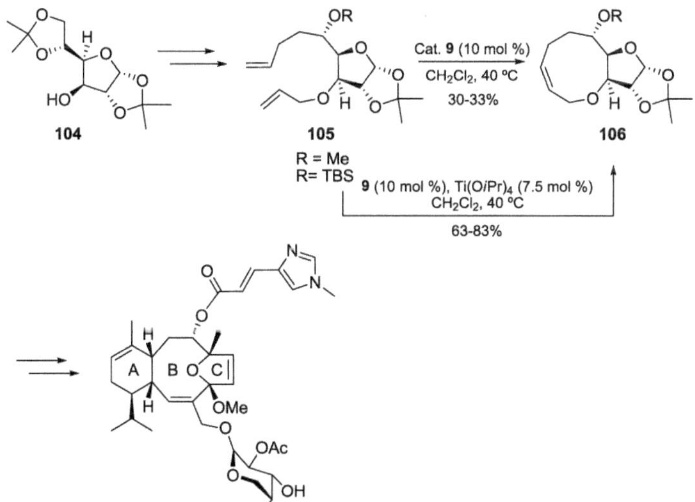

Scheme 19 Synthesis of **106** developed by Kaliappan and Kumar for the preparation of eleutherobin

The related nine-membered oxacycles eicosanoid oxylipins, (−)-halicholactone **114** and (−)-neohalicholactone, were isolated from the marine sponge *Halichondria okadai* (Kadota) collected from Daiozaki, Japan [104]. Halicholactone **114** exhibited weak inhibitory activity against farnesyl protein transferase and lipoxygenase of guinea pig polymorphonuclear leukocytes.

Two synthetic approaches to (−)-halicholactone **114**, involving a RCM as the key step, were reported by Takemoto et al. [105] and Takahashi et al. [106].

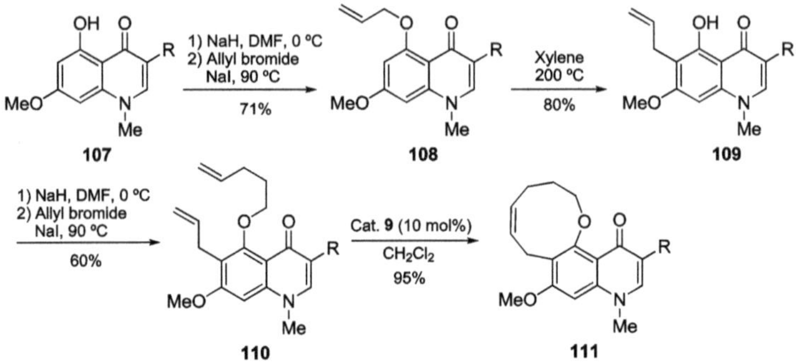

Scheme 20 Synthesis of oxoninoquinolones by RCM

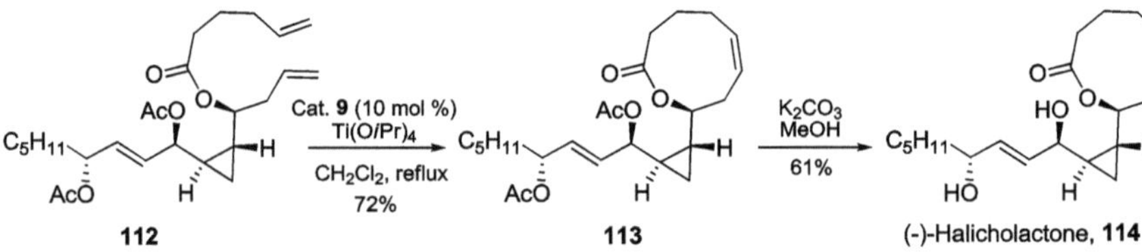

Scheme 21 Synthetic approach to (−)-halicholactone by using a RCM

The authors applied a RCM to dienoic ester **112**, using the first-generation Grubbs catalyst **9** at high dilution in the presence of Ti(O*i*Pr)$_4$ as the Lewis acid, to build the nine-membered structure **113**, which was converted to (−)-**114** by hydrolysis of the acetyl protecting groups (Scheme 21).

Topsentolides are oxylipins which were isolated in 2006 from the marine sponge *Topsentia* sp. [107]. They are characterized by a 20-carbon skeleton and an unsaturated unbranched alkyl side chain containing hydroxy or epoxy groups. They are believed to be formed by lipoxygenation, followed by cyclization of unsaturated fatty acids. They have similar structural frames to halicholactone and neohalicholactone and exhibit moderate cytotoxicity against SK-OV-3 and SK-MEL-2 human solid tumor cell lines. The first total synthesis of (8*S*,11*S*,12*R*)-topsentolide B$_3$ **121** was recently reported [108]. The authors followed a convergent strategy for the stereoselective synthesis of topsentolide B$_3$ (Scheme 22) using the Maruoka asymmetric allylation and a key regioselective RCM. Installation of the chiral 1,2-diol group was achieved through Sharpless asymmetric dihydroxylation by using an AD-mix reagent [109]. To note that the RCM of compound **120**, catalyzed by Grubbs second-generation catalyst (10 mol %) in refluxing dichloromethane, proceeded smoothly and afforded topsentolide B$_3$ in 78 % yield. Although fragment **120** contained two other double bonds in the alicyclic chain, only terminal double bonds were involved selectively to form the nine-membered lactone ring. This result implied that the catalyst did not insert into the double bond at the sterically hindered region and the double bond α to the ester hydroxy group. In this case, the reaction was highly *Z*-selective, and there was no

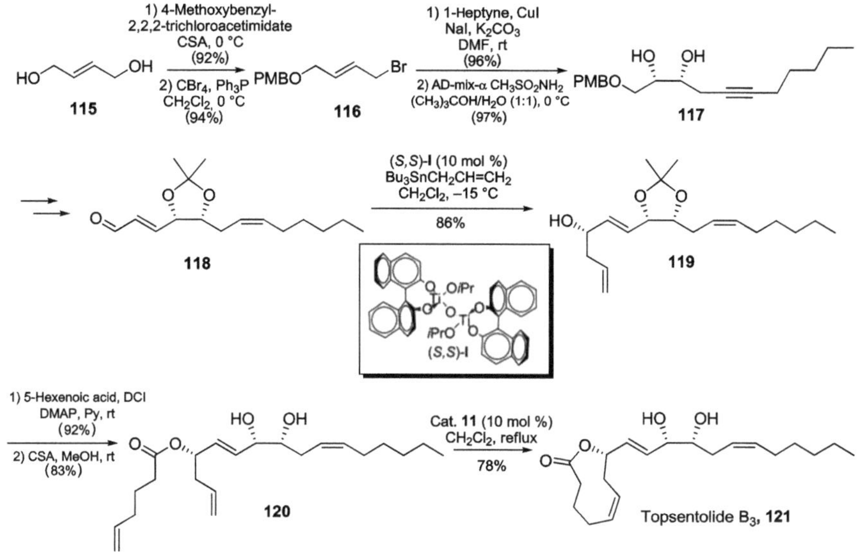

Scheme 22 Stereoselective synthesis of topsentolide B_3, **121**

spectroscopic or chromatographic evidence for the formation of either the *E*-isomer or other side products.

Very recently, Fernandes and Kattanguru have reported the first total synthesis of (8*S*,11*R*,12*R*)-topsentolide B_2. The synthesis features asymmetric dihydroxylation, Wittig and Horner–Wittig olefinations, Roush allylation, and a RCM as the key steps [110]. They observed that the use of Grubbs-II catalyst **11** provided the nine-membered lactone in low yields, whereas the use of Grubbs-I catalyst **9** (10 mol%) and $Ti(O\textit{i}Pr)_4$ (20 mol%) led to improved yields (52 %). Subsequently, the same authors successfully reported a shorter and improved total synthesis of topsentolide B_2 and its C8-epimer in eight steps with better overall yields (about 10 %). The key steps involve a regioselective asymmetric dihydroxylation, a diastereoselective Roush allylation, and a RCM. They have also established the stereochemistry of natural topsentolide B_2 [111]. Very recently, an enantiodivergent approach to enantiopure topsentolide analogs was reported, involving a RCM as the key step with the first-generation Grubbs catalysts **9** [112].

It is worth pointing out that eight- and nine-membered rings are generally obtained with a (*Z*)-endocyclic olefin when a RCM was used to construct such rings. A bibliographic search only provides a few examples describing the formation of an oxonene with an (*E*)-double bond [113], which was observed during the construction of 5-9-5 tricycles through RCM of dienes [114]. The authors observed that for dienes **122**, the relative stereochemistry of the substituents on the five-membered rings influences the olefin geometry significantly, resulting in the synthesis of an unsaturated nine-membered cyclic ether with an (*E*)-configuration for the double bond. Diene **122a** differs from **122b** (Scheme 23) in the stereochemistry

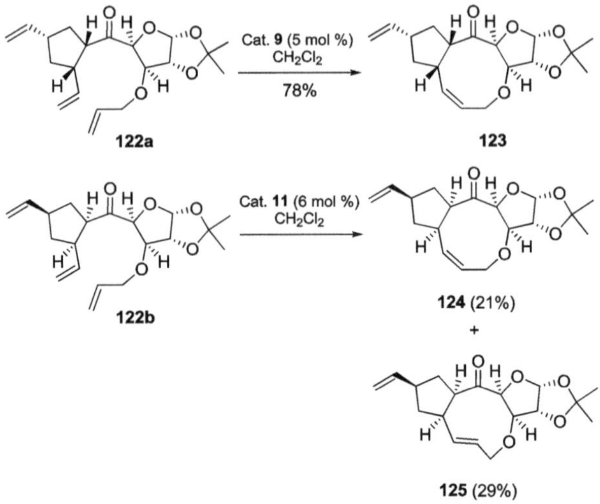

Scheme 23 Effect of the relative stereochemistry of the substituents on the olefin geometry

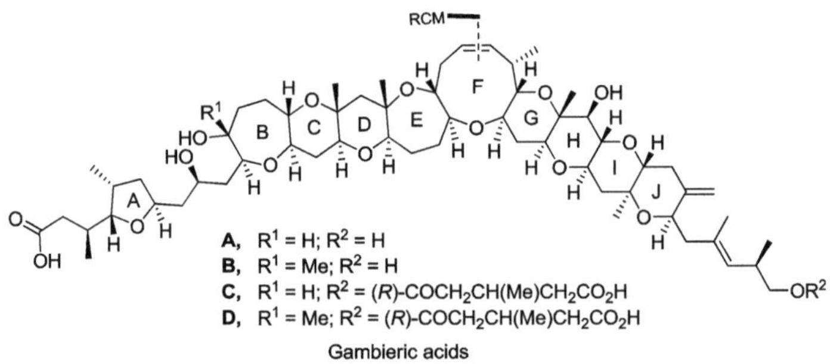

Fig. 5 Gambieric acids

of the substituents of the five-membered rings fused to the nine-membered ring to be formed during RCM. Remarkably this subtle difference leads to the formation of the thermodynamically less stable *E*-alkene in the oxonene. Indeed DFT calculations show that compound **125** is 3.4 kcal/mol more stable than **124**, while **123** is 5.6 kcal/mol more stable than its *trans*-analog. Thus formation of *trans*-oxonene **125** competes favorably and leads to a mixture of *cis*- and *trans*-oxonene **124** and **125**, respectively. It is not the strain energy associated with the incorporation of *trans*-alkene in the ring but probably the overall stability of the resulting ring system that dictates the olefin geometry.

The use of the RCM to build polycyclic ether skeletons is important. Sasaki and Sato described [115] a convergent synthetic route to the CDEFG-ring system **127** of the marine natural products, gambieric acids (Fig. 5), a polycyclic ether skeleton with potent antifungal properties. Thus, the RCM of diene **126** with the second-generation Grubbs catalyst **11** led to the nine-membered F-ring of the desired CDEFG-ring system in 98 % yield (Scheme 24).

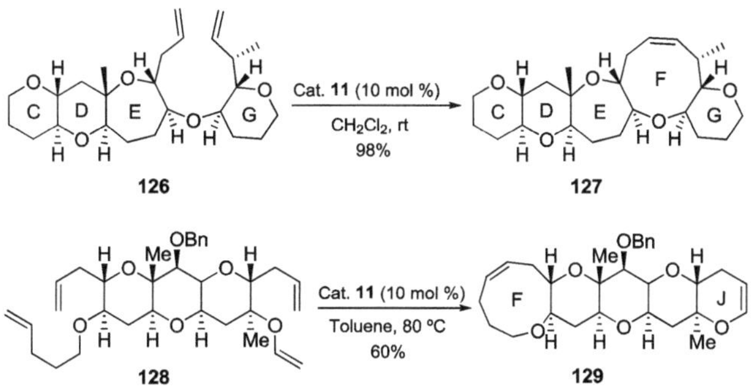

Scheme 24 Use of the RMC to synthesize oxonene rings in polycyclic ether skeletons

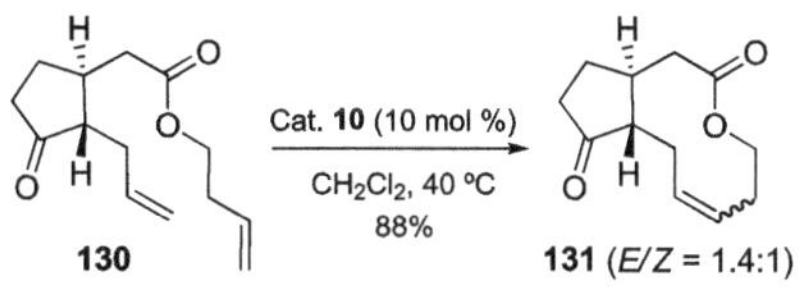

Scheme 25 Fürstner and Müller preparation of ten-membered oxacycles by RCM [118]

Clark et al. reported [116] a rapid two-directional synthesis of the F and J rings (Compound **129**) of gambieric acids by iterative double RCM in which the nine-membered F-ring and the six-membered J-ring were created in one step in 60 % yield. The precursor **128** (Scheme 24) was derived from D-glucal.

Finally, it should be noted that the RCM strategy was successfully utilized for the stereoselective synthesis of the BCDE fragment of brevetoxin A [117].

2.1.3 Synthesis of Ten-Membered Oxacycles

The first construction of a ten-membered lactone using a RCM was reported by Fürstner and Müller in 1997 in their synthesis of the jasmine ketolactone (Z)-**131**, a minor component of the essential oil of jasmine [118]. Heating a dilute solution of **130** (Scheme 25) in the presence of catalyst **10** resulted in the formation of **131** as a mixture of *E*/*Z*-isomers (1.4:1) in 88 % combined yield. Chromatographic separation provided access to the natural product (Z)-**131**. Most likely, the constraint imposed by the five-membered ring favors the cyclization over the dimerization.

Later, Kalesse et al. described, in the synthesis of C7–C17 moiety of epothilones, the preparation of the ten-membered heterocycle **134** by a RCM reaction (Scheme 26) [119]. The role of the substituent during the cyclization is not clear. The stereochemical information of **134** was used to establish another stereocenter by a stereoselective alkylation of the medium-sized ring. They also found that the solvent has a significant influence on the *E*/*Z* selectivity.

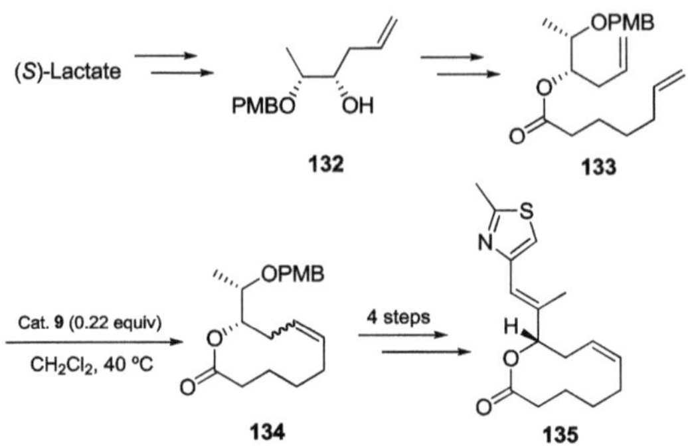

Solvent	Z/E 134
CH_2Cl_2	12:1 (63%)
Toluene	1:1 (59%)

Scheme 26 Synthesis of the C1–C7 fragment of epothilones by RCM

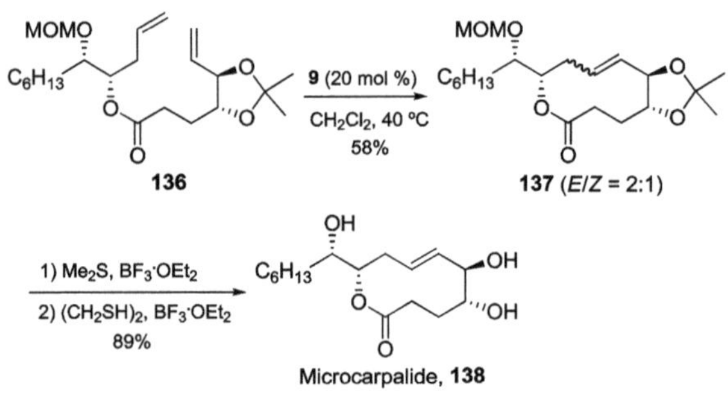

Scheme 27 Total synthesis of microcarpalide **138**

Carda et al. reported a total synthesis of microcarpalide **138** (Scheme 27) [120], a naturally occurring nonenolide with cytotoxic and antimicrofilament activity. This study also featured a RCM as the key reaction. Thus, cyclization of diene **136**, prepared from (*S*,*S*)-tartaric acid and (*R*)-glycidol, with catalyst **9**, led to a 2:1 *E*/*Z* mixture for macrocycles **137**, from which the required *E*-isomer was separated. On the other hand, treatment of compound **136** with the second-generation catalyst **11** furnished almost exclusively the thermodynamically more stable (*Z*)-**137** isomer. These observations are in agreement with those of Grubbs, who found that the *E*/*Z* ratio in ring closure using catalyst **11** is not kinetically controlled, but it is rather the result of an equilibration of the products [121]. The synthesis of **138** was then completed by a sequential deprotection of intermediate **137**.

In the synthesis of herbarumin I, Fürstner et al. [122] revealed several salient features for the RCM of ester-tethered dienes. Diene **139** (Scheme 28) was prepared from D-ribose using conventional transformations. The diol protecting group was an isopropylidene group with the expectation that it would help to stabilize one conformation of **139** that would favor the ring closure. Semiempirical

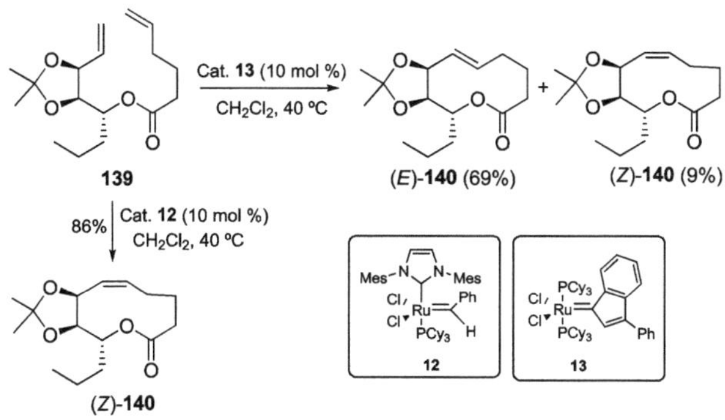

Scheme 28 Effect of the catalyst on the olefin geometry in the RCM [122]

calculations on **140** revealed that the (*Z*)-isomer is more stable than the (*E*)-isomer (by 3.5 kcal/mol). Therefore, conducting its RCM (or any other appropriate diene) under thermodynamic control, would be expected to be counterproductive for obtaining the (*E*)-alkene present in the natural product. This, in turn, suggested that a RCM catalyst known to equilibrate the initial products should not be employed. Gratifyingly, by using two different RCM catalysts the (*E*)- and (*Z*)-isomers were obtained selectively. Thus, cyclization of **139** with the second-generation Grubbs catalyst **12**, which was known to provide mixtures enriched in the thermodynamically favored product, led to the selective formation of (*Z*)-**140**. In contrast, exposure of **139** to catalytic amounts of ruthenium indenylidene complex **13** afforded the desired (*E*)-**140** as the major product, as well as a small amount of the (*Z*)-isomer.

Following the hypothesis of the conformational constraint, Kobayashi et al. reported the first total synthesis of the proposed structure of diversifolin [123]. The synthesis involves a RCM to give a ten-membered cycle as the key intermediate. Wang et al. reported the synthesis of substituted dibenzo-fused ten-membered cyclic ethers from 2-allylbenzyl alcohols and 2-allylphenols derived from isovanillin, by using a RCM approach to give the desired cyclic ethers [124].

Fürstner and Schlede [125] reported a concise synthesis of ten-membered rings (Scheme 29), which constitutes the key intermediates in the synthesis of the marine natural product ascidiatrienolide A. The *E*/*Z* ratio obtained in the RCM reaction is found to be dependent on the relative configuration of the substituents of the precursor of the cyclization as well as on the catalyst used. Specifically, the ruthenium indenylidene complex **13** and the second-generation Grubbs-type catalyst **12** bearing an *N*-heterocyclic carbene ligand led to the opposite stereochemical results when applied to the *syn*-configured diene **141**, and with an identical outcome with the *anti*-configured analog **143**.

To end this section, Takahashi et al. have recently reported the total synthesis and the structural revision of phomopsin B, a novel polyketide possessing a ten-membered cyclic ether [126]. The total synthesis of the target was achieved by using an intramolecular olefin metathesis as the key step.

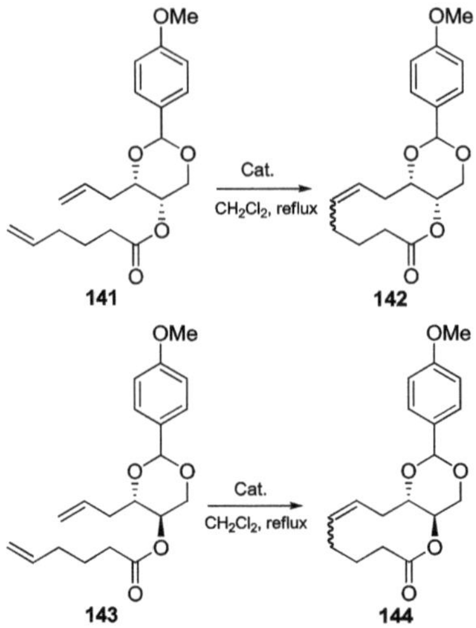

	Cat. **13** (20 mol %)	Cat. **11** (2 mol %)
Yield	83%	79%
E/Z	3.5 : 1	1 : 2.8

	Cat. **13** (20 mol %)	Cat. **11** (2 mol %)
Yield	67%	71%
E/Z	8 : 1	8.4 : 1

Scheme 29 Effect of the catalyst on the olefin geometry in the RCM reported by Fürstner et al.

In summary, as it has been related above, the increasing use of RCM to achieve the medium-ring oxacycles must be highlighted.

2.2 *Cyclization to Form Carbon–Carbon Single Bond*

This strategy has been widely utilized for the construction of medium-sized ethers. This section is focused on metal-catalyzed cyclization, free radical cyclization, and alkylations.

2.2.1 Free Radical Cyclization

Free radical cyclizations are important synthetic tools due to mild reaction conditions, along with high levels of chemoselectivity and stereoselectivity. However, there are problems associated with the formation of medium-sized rings. Thus, there are only few examples in the literature for the construction of eight-membered ring systems, in general, and ethers, in particular, by radical cyclization. Herein, recent advances in radical cyclizations that have led to the development of new methods to efficiently construct eight-membered ethers are reported. To the best of our knowledge, up to now nine- and ten-membered ring ethers have not been synthesized by using radicals.

Majumdar [127] reported thiophenol-mediated radical cyclization reactions to form benzoxocine derivatives. As shown in Scheme 30, the eight-membered ether **148** was formed from enyne **145** by using a regioselective thio-radical addition

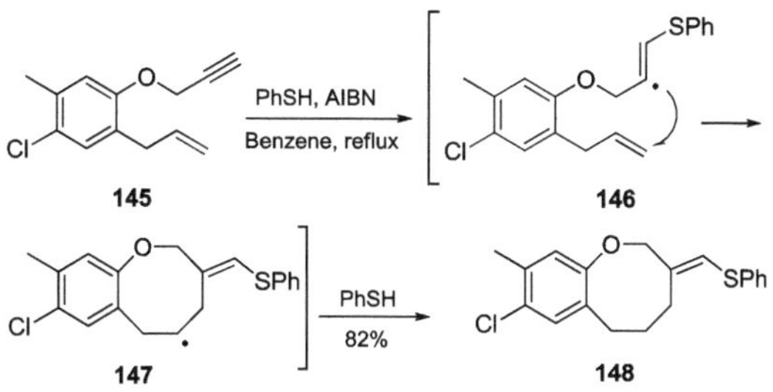

Scheme 30 Thiophenol-mediated radical cyclization reported by Majumdar et al.

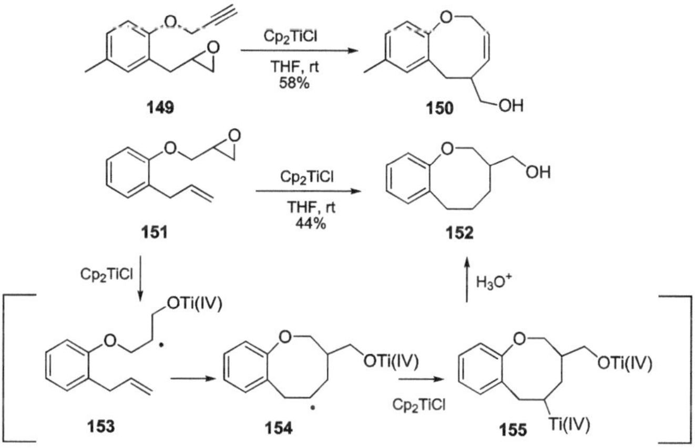

Scheme 31 Radical cyclization reactions reported by Roy et al.

followed by an intramolecular 8-*endo-trig* radical cyclization. The planar and rigid molecular scaffold provided by the bicyclic framework in **145** limits the conformational degrees of freedom of the alkyne and alkene side chains so that a highly and efficient unusual 8-*endo* vinyl radical cyclization can occur.

This regioselectivity is in line with that found in the existing literature, which clearly demonstrates the predominance of the 8-*endo* cyclization mode over the 7-*exo* mode [128].

Mandal and Roy [129] developed novel titanocene chloride-mediated 8-*endo* radical cyclizations (Scheme 31). Through the generation of radicals from epoxides **149** and **151**, a cyclization took place to form the eight-membered ring ethers **150** and **152**, respectively. The mechanism for the formation of **152** from **151** is depicted in Scheme 31. Ti(III) reagents reacted as a single electron reductor to open the epoxide motif in **151** to form intermediate **153**, which then underwent an 8-*endo* radical cyclization to form **154**. Radical intermediate **154** was reduced by a second equivalent of Ti(III) species to form **155**, which underwent protonation to furnish **152**.

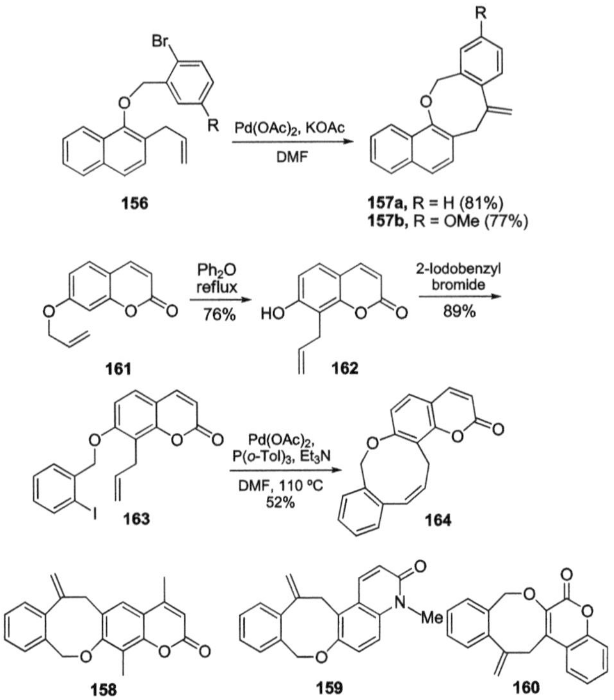

Scheme 32 Intramolecular Heck reaction developed by Söderberg et al., Majumdar et al., and Chattopadhyay et al.

2.2.2 Metal-Mediated Cyclization

This method has been mainly applied with success for the synthesis of eight- and nine-membered ethers. We will describe the syntheses of both sizes of cycles in a common subsection. On the other hand, to our knowledge, this approach has not been applied for the construction of ten-membered ring ethers.

Palladium-catalyzed cyclization implies the addition of the carbon–palladium bond to another carbon-based center. Provided with easily accessible starting materials and mild reaction conditions, carbopalladation has been viewed as one of the most important methods in organic synthesis [130]. Söderberg et al. [131], Majumdar et al. [132], and Chattopadhyay et al. [133] reported the synthesis of medium-sized oxa-heterocycles by palladium-catalyzed intramolecular Heck reaction. As shown in Scheme 32, eight-membered ethers **157** were obtained from the corresponding aryl bromide precursors **156** in good yields. Similar reactions were also reported by Chattopadhyay et al. who described a sequential Claisen rearrangement and intramolecular Heck cyclization to access benzoxocine- and benzoxonine-fused coumarin and quinolone derivatives, such as **158–160**, in moderate yields (Scheme 32) [133]. Thermally induced Claisen rearrangement of 7-allyloxycoumarin **161** gave **162**, which upon alkylation with 2-iodobenzyl bromide gave the Heck precursor **163**. Standard Heck conditions gave the 9-*endo*

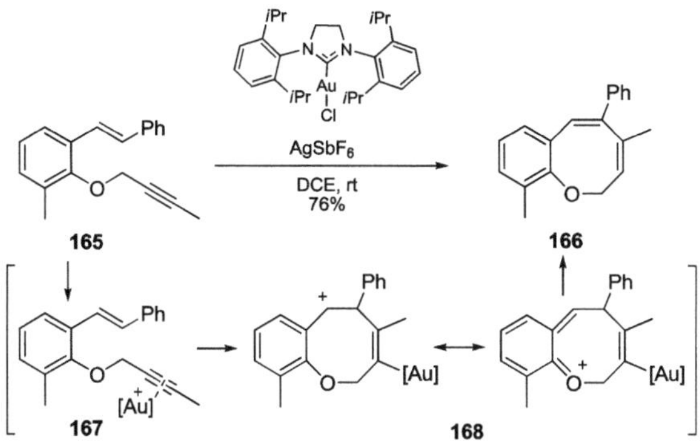

Scheme 33 Au(I) mediated formation of benzoxocines

cyclized product **164** in reasonable yield. Intriguingly, changing the substitution pattern of the linear precursor modified the mode of cyclization. Thus, treatment of 4-allylcoumarin under identical reaction conditions gave the 8-*exo* cyclized product **160**, a result that has to be explained mechanistically [134].

Kumar et al. and Waldman et al. reported the preparation of benzoxocines by using an Au(I)-catalyzed 8-*endo-dig* cyclization (Scheme 33) [135]. Treatment of propargyl ether **165** with an in situ generated cationic Au(I) species yielded benzoxocine **166**. This reaction presumably proceeds through the initial formation of the gold–alkyne complex **167**, followed by an 8-*endo-dig* cyclization to generate **168** which led to the observed product **166**.

2.2.3 Alkylation of Anion

As the previously described approaches, intramolecular enolate alkylation has not found widespread use for the preparation of medium-ring ethers, except for a few eight- and nine-membered ring ethers.

Kim et al. have developed an intramolecular amide enolate alkylation as a key step for the syntheses of several natural products, such as cladiellin diterpene [136], and laurencin [59]. These authors found that the cyclization of (*E*)-allylic chloride amides through an S_N2' pathway furnishes tetrahydropyrans such as **170** (Scheme 34), but when the corresponding (*Z*)-allylic chloride was employed, Δ^4-oxocine **171** was found to be the major product from an S_N2 pathway with only a minor amount of the expected S_N2' product **170**. This result highlighted the key role played by the alkene geometry in directing the course of the cyclization.

Kim et al. used this strategy to achieve the first total synthesis of (*E*)-cladiellin and a variety of 2,11-cyclized cembranoids [137]. Enolization of highly functionalized amide **172** with LiHMDS cleanly afforded the nine-membered ring ether **173** (Scheme 35), presumably through a lithium-chelated (*E*)-enolate.

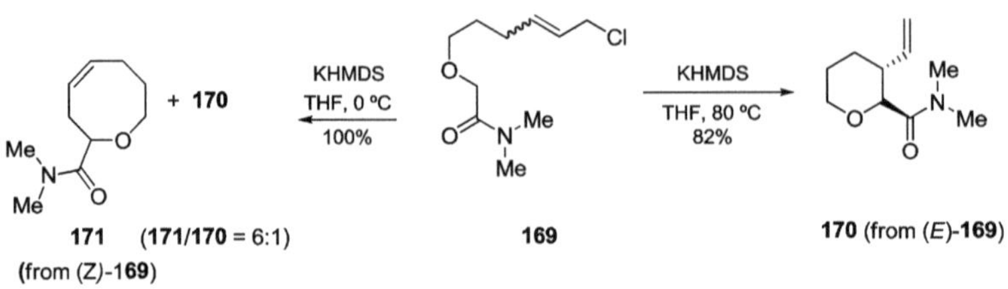

Scheme 34 Strategy of intramolecular amide enolate alkylation developed by Kim et al.

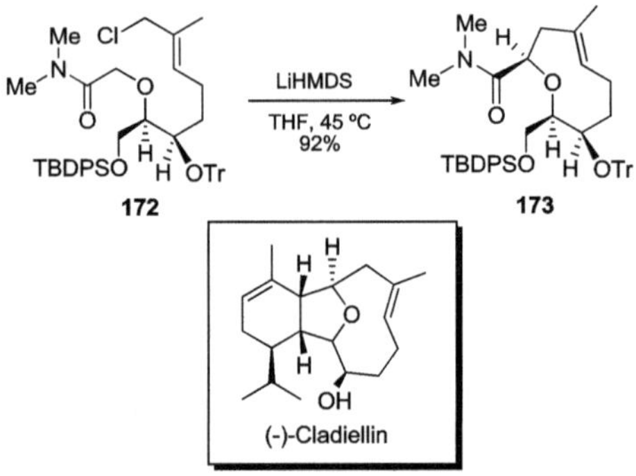

Scheme 35 Application of the Kim et al. method to the total synthesis of (*E*)-cladiellin

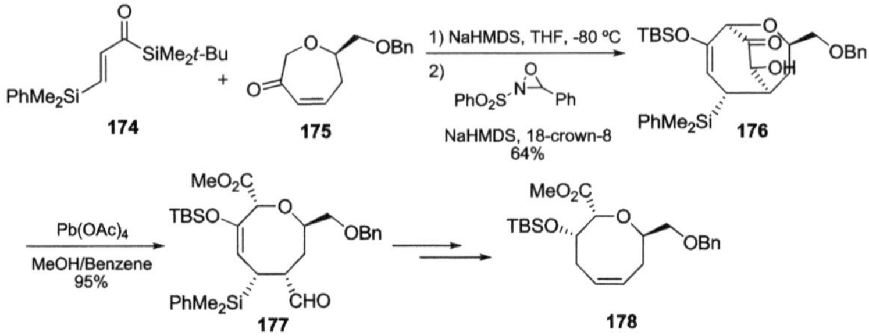

Scheme 36 One-pot [3+4]-annulation/oxidation reaction reported by Takeda et al.

Takeda et al. reported the formation of medium-sized carbo- and oxacycles by using a Brook rearrangement-mediated [3+4]-annulation reaction [138]. This method was then employed in a formal synthesis of (+)-laurallene [139] and (±)-prelaureatin [140]. As shown in Scheme 36, for the preparation of (+)-laurallene, acryloyl silane **174** and the unsaturated seven-membered ketone **175** were converted, using Davis' oxaziridine, to **176** by a one-pot [3+4]-annulation/oxidation sequence. Bicyclic intermediate **176** was then submitted to an oxidative cleavage producing the eight-membered ether **177**, which was further transformed to **178**, an intermediate in the Crimmins' synthesis of (+)-laurallene.

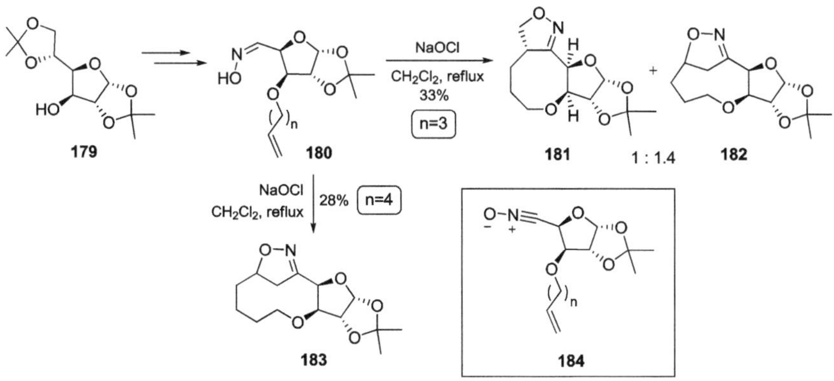

Scheme 37 Intramolecular nitrile oxide–alkene cycloaddition

2.2.4 Cycloaddition and Pericyclic Reaction

Shing et al. reported the synthesis of medium-sized ethers from carbohydrates by utilizing intramolecular 1,3-dipolar cycloaddition of nitrile oxides and alkenes [141]. As shown in Scheme 37, on treatment with NaOCl, oximes **180** ($n = 3, 4$) were converted into the corresponding medium-sized membered ethers **181**–**183** (yields around 30 %) via the nitrile oxide intermediates **184**. The results reported indicated that the regioselectivity of the intramolecular 1,3-dipolar cycloaddition of these nitrile oxides is dependent on the length of the alkenyl chain.

The ring size represents a challenge for the use of pericyclic and cycloaddition approaches in the synthesis of medium-sized ethers. Up to now, only a limited number of reports have detailed the successful use of [8+2]-cycloaddition processes in the formation of ten-membered ring systems. Herndon et al. have reported the synthesis of various oxo-bridged ten membered ring systems [142] using simple dienylfurans as 8π-components such as **185** (Scheme 38). Reaction of [8+2]-cycloadducts with electrophilic reagents occurred selectively at the bridgehead double bond. In mechanistic studies on related systems, it has been found that these reactions likely proceed through initial [4+2]-cycloaddition to afford oxacycle **186**, which then undergoes a [1,5]-vinyl shift, generating the observed product **187** [143].

Holmes et al. applied a Claisen rearrangement to construct the eight-membered ether motif in their total synthesis of (+)-laurencin [55]. Thus, selective reduction of maleic diester **188**, followed by a protection as an acetonide and partial reduction and addition of vinylmagnesium bromide in the presence of cerium(III) chloride, provided allylic alcohol **189** as a 1:1 mixture of diastereoisomers. Then, acetonide removal followed by selective protection and treatment with phenylselenoacetaldehyde diethyl acetal afforded the dioxane **190** as a mixture of diastereoisomers, which was then oxidized to give the selenoxide and then treated with DBU in refluxing xylene to provide the Claisen rearrangement product **191** in 73 % yield (Scheme 39). This method allows a rapid entry into natural product cores as

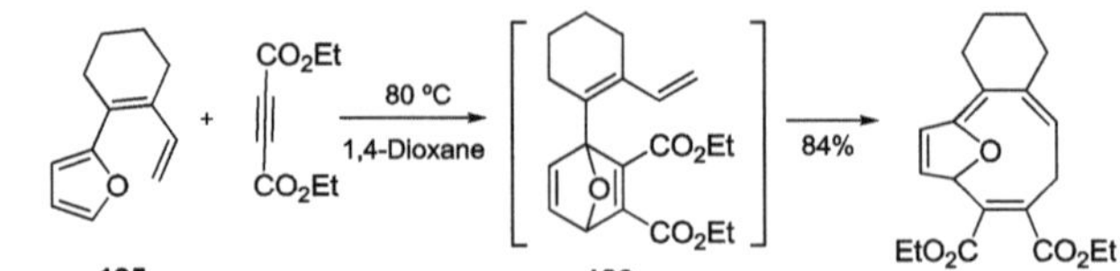

185 186 187

Scheme 38 [8+2]-Cycloaddition reaction to form ten-membered rings

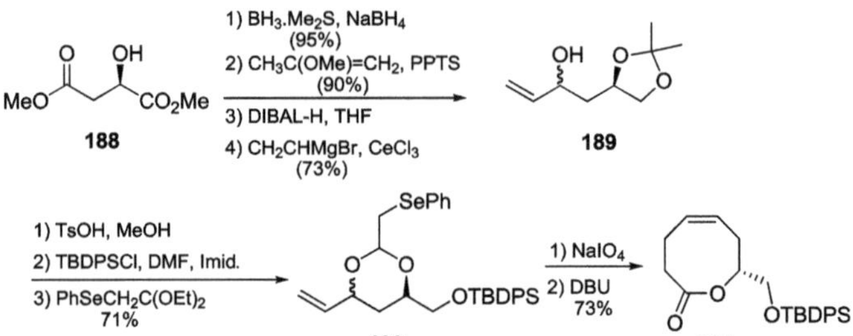

188 189 190 191

Scheme 39 Claisen rearrangement to construct the eight-membered ether motif in the synthesis of (+)-laurencin by Holmes et al.

Holmes et al. have later demonstrated in their synthesis of the nine-membered ether (+)-obtusenyne **6** [144] and the core of eunicellin **7** [145].

2.2.5 Cyclization by Prins-Type Reaction

Overman et al. developed a powerful method for the stereoselective synthesis of medium-ring ethers by Prins cyclization, which resulted in a *cis* orientation of the side chains adjacent to the ether oxygen via an intramolecular ene reaction mechanism [146].

The total synthesis of (+)-laurencin was efficiently achieved by using this method [53] (Scheme 40). The synthesis started by modified Brown's asymmetric allylation using allyl SEM ether **192** and propionaldehyde. After silylation of the ensuing alcohol, the resulting species **193** was subjected to a Suzuki coupling with vinyl thiobromide **194** followed by cleavage of the TBS group to give **195**. Acetalization of compound **195** with the bromoether **196** and subsequent protective group replacement afforded the cyclization precursor **197**. The Prins cyclization of **197**, which would proceed via the oxocarbenium intermediate **198**, efficiently produced a Δ^4-oxocene **199** with the 2,8-*cis* side chains. Subsequent manipulation, involving oxidation, reduction, and Wittig reaction, among other reactions, furnished (+)-laurencin **2**.

This method was also efficiently employed for the construction of the eight-membered cyclic ether core of (–)-laurenyne [52].

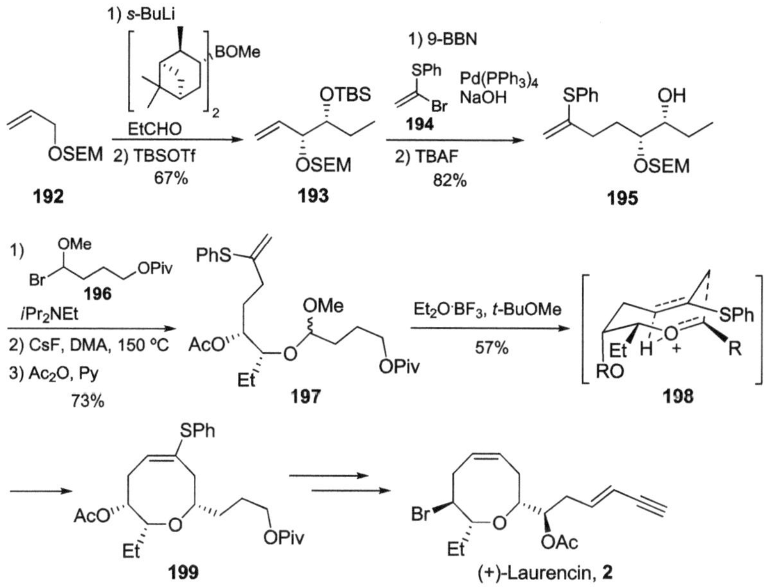

Scheme 40 Utilization of the Prins cyclization to synthesize (+)-laurencin

2.3 Cyclization to Form Carbon–Oxygen Single Bond

An obvious synthetic approach for the construction of cyclic ethers is the intramolecular cyclization of linear precursors through a C–O bond forming event. Despite this method exhibits evident kinetic and thermodynamic difficulties for medium-sized ring ethers, numerous successful approaches have been reported.

2.3.1 Intramolecular Cyclization of Hydroxy Ketones and Thioketals

Reductive cyclization of hydroxy ketones has been used frequently during the total synthesis of natural products possessing medium-sized ether moieties. Kumar et al., following the pioneering work by Nicolaou et al. on the synthesis of the marine ladder polyethers [147], reported the total synthesis of (−)-*cis*-lauthisan [148]. As shown in Scheme 41, treatment of the hydroxy ketone **200** with Et_3SiH and TMSOTf resulted exclusively in the *cis*-disubstituted eight-membered cyclic ether **201**. Carreño et al. also reported the total synthesis of (−)- and (+)-*cis*-lauthisan using the same strategy [149]. Similarly, **202** underwent reductive cyclization to afford the same product **201** also in moderate yield. The formation of diastereomers was probably due to the pseudoaxial delivery of hydride to the oxocarbenium ion intermediate.

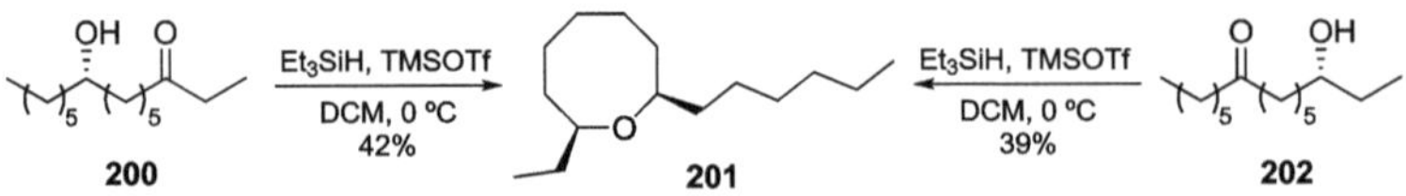

Scheme 41 Kumar et al. and Carreño et al. reductive cyclization of hydroxy ketones

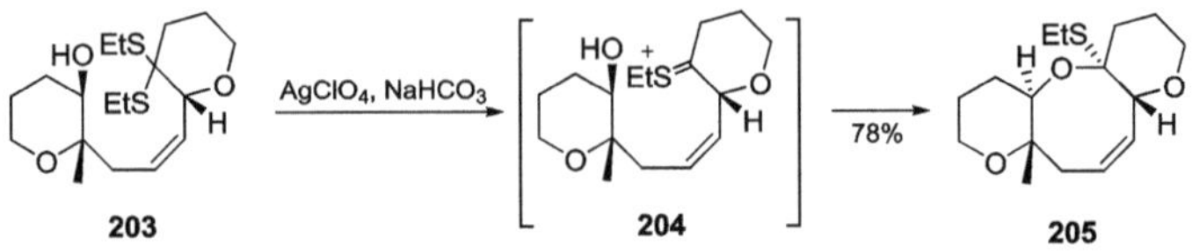

Scheme 42 Mild reductive cyclization developed by Nicolau et al. and applied to the synthesis of brevetoxins A and B

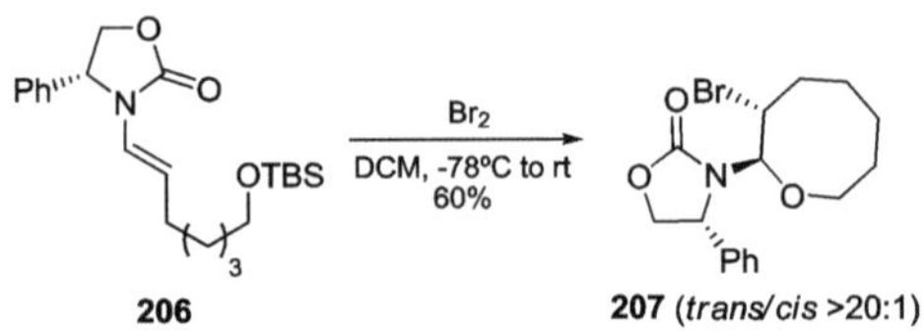

Scheme 43 Halo-etherification of chiral enamides to form halogen-containing cyclic ethers

Nicolaou et al. applied a milder version of this reductive cyclization, in the form of a hydroxyethyl thioketal reductive cyclization [150], during the total synthesis of brevetoxins A [151] and B [152, 153]. Treatment of **203** with silver perchlorate led to an oxocene in high yield, despite the strain and entropic barrier encountered during the cyclization. In this reaction, it is presumed that the role of $AgClO_4$ on **203** induces the formation of a transitory hydroxy thionium ion **204**, a highly reactive species that undergoes facile ring closure to oxocene **205** (Scheme 42). The (*Z*)-double bond of **203** plays a crucial role in the cyclization event by reducing rotations; in the absence of a (*Z*)-double bond, the ring closure does not occur. Accordingly, this strategy is usually preceded by a Wittig reaction.

2.3.2 Halo-Etherification of Enamide

Hsung et al. reported a stereoselective halo-etherification of chiral enamides to obtain halogen-containing cyclic ethers [154]. As shown in Scheme 43, treatment of the chiral enamide **206** with bromine furnished the eight-membered ether **207** in good yield with the *trans*-product as the major diastereomer. Ten-membered ring ethers could also be constructed, although yields are not good.

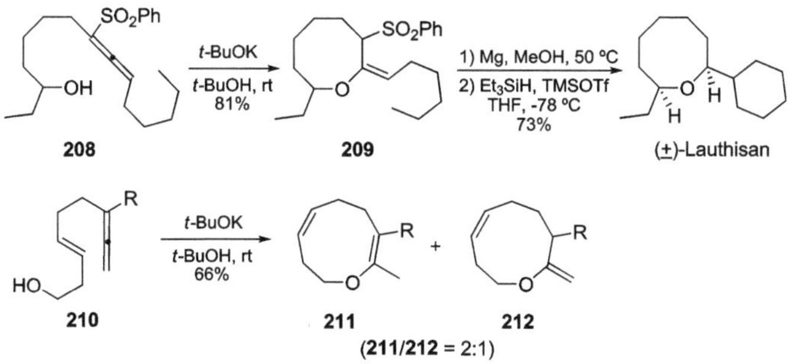

Scheme 44 Formation of eight- and nine-membered cyclic ethers by intramolecular nucleophilic addition of a hydroxy group to an allene

2.3.3 Intramolecular Cyclization of Hydroxyl Allene

The intramolecular nucleophilic addition of a hydroxyl group to an allene substituted by an electron-withdrawing phenylsulfonyl group was employed in the synthesis of the eight-membered oxacycle of (±)-lauthisan [155] (Scheme 44). Exposure of the allene **208** to *t*-BuOK in *t*-BuOH at room temperature for 40 min furnished (*E*)-8-ethyl-2-hexylidene-3-(phenylsulfonyl) oxocane **209** in 81 % yield. Dephenylsulfonylation with Mg in MeOH followed by the reduction of the enol ether moiety with Et_3SiH afforded (±)-lauthisan in 73 % yield.

These basic conditions also allowed the rapid *endo*-mode ring closure of the allene derivatives **210** to furnish 2,3,6,7-tetrahydro-9-methyloxonines **211** in good yield as single isomers [156]. In the case of sulfonyl derivative **210** (R = SO_2Ph), the *endo*-mode reaction proceeded as expected to give the cyclic products in 66 % yield as a mixture of **211** and its isomer **212** with an *exo*-methylene moiety in a ratio of ca. 2:1.

2.3.4 Metal-Mediated Heterocyclization

Hanaoka et al. developed a method for the preparation of oxocane as well as the oxonane skeletons by taking advantage of the inherent property of the alkyne–$Co_2(CO)_6$ complexes, easily prepared by the addition of $Co_2(CO)_8$ to propynyl alcohols or ethers. These complexes led, upon treatment with a Lewis acid, to the corresponding propynyl cation species, which were subsequently captured by various nucleophiles (Nicholas reaction). This method is a powerful tool for the preparation of medium-sized oxygen atom-containing heterocycles [157] (Scheme 45). For instance, the oxonane **217** was synthesized following this approach from propargylic alcohol **216**.

Martin et al. also reported the synthesis of cyclic ethers based on intramolecular Nicholas reaction. Thus, they focused on the intramolecular attack of a hydroxy

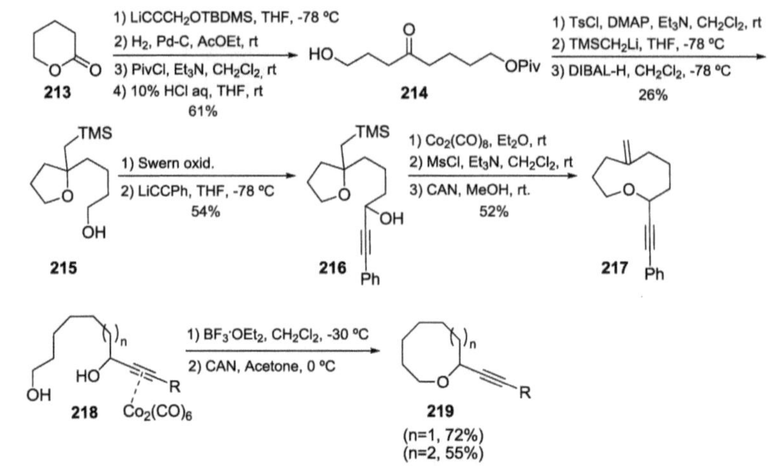

Scheme 45 Cobalt-catalyzed heterocyclizations to form medium-sized ethers

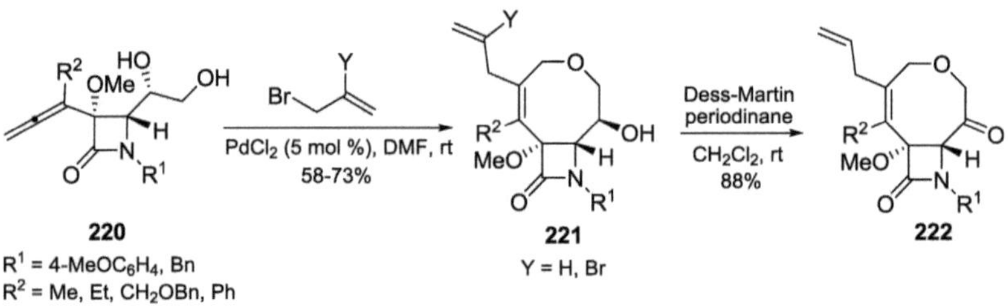

Scheme 46 Formation of *trans*-fused β-lactam oxocines via Pd(II)-mediated heterocyclization

group on a carbocation generated by acid treatment of *exo*-$Co_2(CO)_6$-complexed propargylic alcohols **218** as a method for the preparation of cyclic ethers (Scheme 45). The practical importance of this method remains in the facile introduction and removal of the cobalt moiety, the extraordinary stabilization of the propargylic carbocation, and the impact of a number of different factors on the stereochemistry of the newly created chiral center, as well as on the variety of synthetic transformations allowed by the presence of the triple bond [158].

Alcaide et al. reported the formation of *trans*-fused β-lactam oxocines **221** in good yields in a totally chemo- and regioselective fashion using the $PdCl_2$-catalyzed cyclization of α,β-allene diols **220** with allyl halides (Scheme 46) through an 8-*endo*-*trig* process by attack of the primary hydroxy group to the terminal allene carbon, followed by a Heck-type migratory alkene insertion and halide elimination [159].

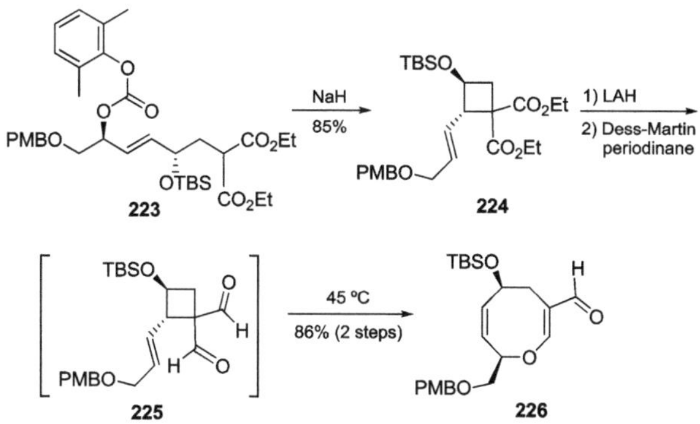

Scheme 47 Formation of dihydrooxocines via retro Claisen rearrangement

2.3.5 Pericyclic Reaction

During the synthesis of (+)-laurenyne, Boeckman et al. have reported the use of a retro-Claisen transformation for the formation of the core of dihydrooxocine **226** [160] (Scheme 47). Thus, cyclization of allylic carbonate **223** via an S_N2' pathway leads to the diastereoselective formation of cyclobutane **224**, which after a reductio-n–oxidation sequence of the geminal esters provides the required substrate **225**. Subsequent heating initiates the thermal retro-Claisen rearrangement, which affords the desired dihydrooxocine–aldehyde **226** in high yield.

Suzuki et al. reported an electrocyclic ring-opening/ring-closing cascade for the generation of 2-benzoxocin derivatives (Scheme 48) [161]. These authors reported that heating a toluene solution of 1-acyloxybenzocyclobutene **227** resulted in a mixture of 2-benzoxocin **228** and naphthalene **229**. This reaction presumably proceeds via an initial retro-4π-electrocyclization of **227** to give quinodimethane intermediate **230**. Direct oxo-8π-electrocyclization of **230** led to **228**, while competing 6π-electrocyclization afforded dihydronaphthalene **231** as a precursor of the isolated naphthalene **229**.

2.4 Ring Expansion and Rearrangement

Ring expansions allow the transformation of more readily available small rings into less common medium-sized ethers which provide an alternative way to construct the medium-sized ethers as compared to direct ways. Among the advantages of this approach, the ring expansion of smaller rings circumvents many of the entropic penalties implied in the formation of the ring and overcomes the need for high-dilution conditions. These processes often occur from a preformed bicyclic system or through the transient formation of a polycyclic intermediate, which produces the desired heterocycle.

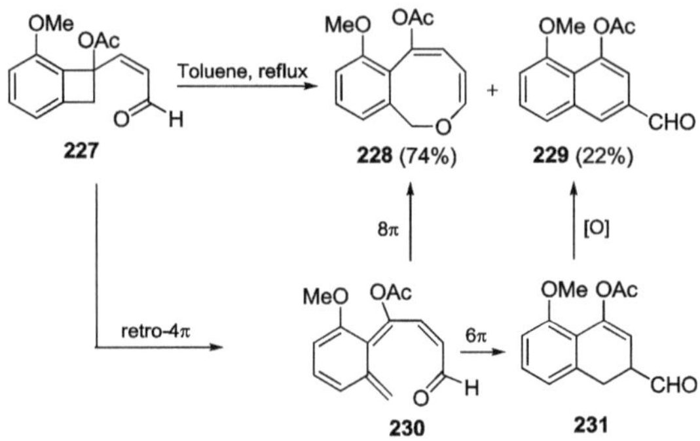

Scheme 48 Electrocyclic ring-opening/ring-closing cascade for the generation of 2-benzoxocin derivatives

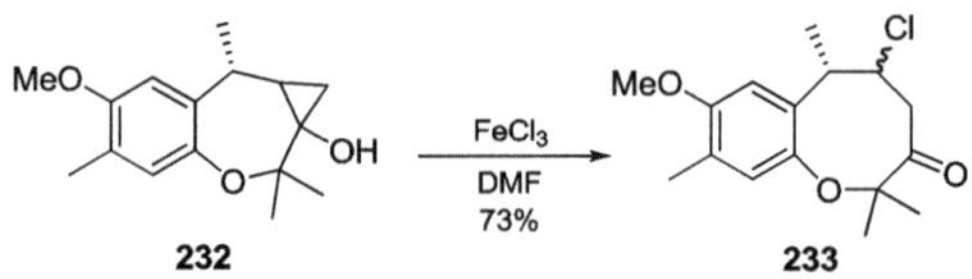

Scheme 49 $FeCl_3$-promoted ring expansion to form eight-membered ethers

2.4.1 Ring Expansion of Cyclopropanes

Ollivier et al. reported a $FeCl_3$-promoted ring expansion to form eight-membered ethers [162]. As shown in Scheme 49, the cyclopropyl alcohol **232** was treated with $FeCl_3$ to provide the chlorinated eight-membered ether **233** in good yield.

Venkateswaran et al. developed a novel construction of the benzoxocane ring system present in the sesquiterpenes heliannuol A and K and helianane in which a ring enlargement, involving a selective central bond cleavage in a cyclopropane annulated cycloheptane ring, is the key step [163]. Thus, the cyclopropyl ketone **235**, formed by the treatment of benzoxopinenone **234** with diazomethane in the presence of a catalytic amount of palladium acetate, was subjected to a catalytic hydrogenation, and the expected selective bond fission was observed, delivering the benzoxocanone **236** exclusively in excellent yield. Reduction of **236** with sodium borohydride in methanol furnished 5-deoxyheliannuol A **237** in a stereocontrolled manner (Scheme 50).

2.4.2 Ring Expansion of Five-Membered Ethers

Nakata et al. reported the stereoselective 1,4-rearrangement ring expansion of tetrahydrofurans via bicyclo[3.3.0]oxonium ions **239** to synthesize oxocanes [164]. The use of a mesylate as an efficient leaving group was an advantage for this ring expansion. As shown in Scheme 51, treatment of tetrahydrofuran **238** with

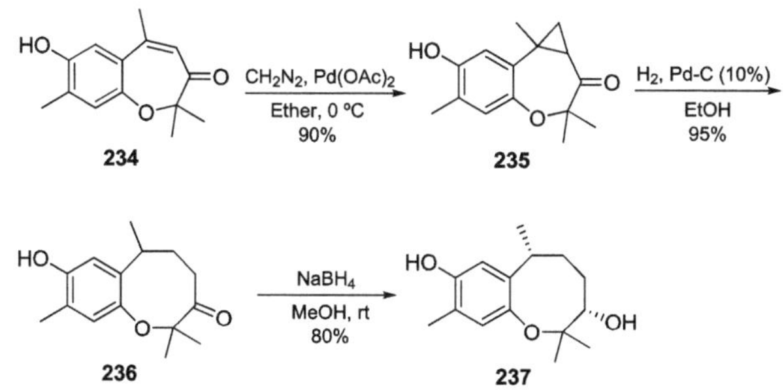

Scheme 50 Construction of the benzoxocane ring system by a selective central bond cleavage of a cyclopropane ring

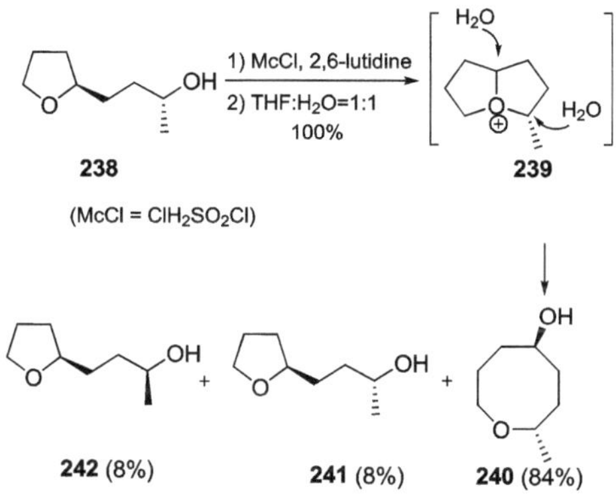

Scheme 51 Formation of oxocines by ring expansion of five-membered ethers

chloromethanesulfonyl chloride (McCl) and a base generated oxocane **240** and tetrahydrofurans **241** and **242**. Presumably, the reaction proceeds via the formation of the *meso*-bicyclo[3.3.0]oxonium ion **239** through displacement of the leaving group by the tetrahydrofuran oxygen. The stereochemically well-defined nature of bicyclic compound **239** was apparently transferred to the products, as only one single diastereoisomer of **240** was obtained. The results with diastereomeric pairs revealed that the present rearrangement ring expansion takes place stereoselectively and stereospecifically.

Recently, Snyder et al. have disclosed a novel procedure for bromonium-induced ring expansion of tetrahydrofurans effected by bromodiethylsulfonium bromopentachloroantimonate (BDSB), affording medium-sized cyclic bromoethers resembling those of the *Laurencia* C15 acetogenins. This process is fast, regio- and stereoselective and has been demonstrated to produce diverse stereochemically eight-membered bromoethers and nine-membered derivatives [165]. This reaction was initiated by the formation of the bromonium ion **244** (Scheme 52). The more reactive bromonium diastereoisomer was opened according to a 5-*exo* attack of the

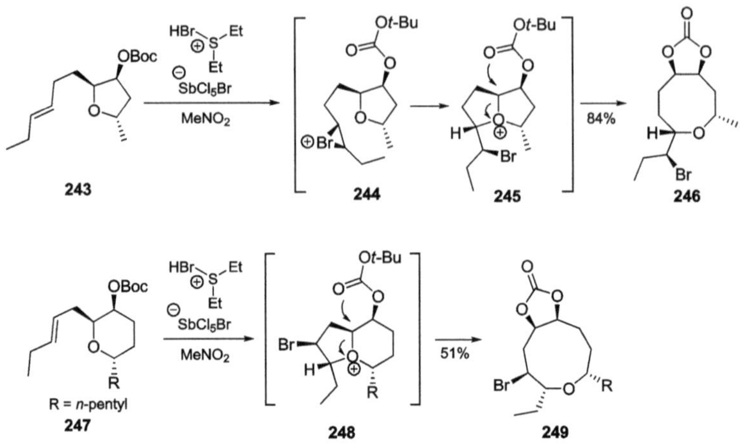

Scheme 52 Bromonium-induced ring expansion to afford eight- and nine-membered oxacycles, by Snyder et al.

tetrahydrofuran oxygen which generates the intermediate bicyclic oxonium ion **245**. The bicyclic compound **245** may then undergo a regioselective ring opening by the nucleophilic attack of the neighboring carbonate moiety producing the observed oxocane **246**. The authors applied this strategy to the racemic formal total syntheses of laurefucin and (±)-*E*- and (±)-*Z*-pinnatifidenyne [166].

A closely related approach has been described by Li et al. who have disclosed an enantioselective synthesis of the Δ^4-oxocene core present in (+)-laurencin, (+)-prelaureatin, and other members of the laurencin family. The synthesis was accomplished in eight steps by novel one-pot regio- and stereoselective ring cyclization/fragmentation/expansion cascade from the tetrahydrofuran precursors, which were prepared by the cyclization of vinylsilanes [167]. This process is highlighted by an intramolecular oxo-carbenoid insertion and a β-silyl fragmentation sequence. Thus, treating aldehyde **250** with a slight excess of ethyl diazoacetate (Scheme 53) in the presence of a catalytic amount of anhydrous $SnCl_2$ in CH_2Cl_2 at room temperature cleanly afforded the β-keto ester **251** (62–70 %), accompanied with Δ^4-oxocene **252** (20–25 %) as a single diastereomer. The authors proposed that **252** was generated via a tricyclic oxo-ylide **255**, stereoselectively formed through an oxo-carbenoid resulting from the insertion of the oxygen atom of the tetrahydrofuran ring into the generated carbene. In this highly organized hydrindane-type intermediate **255**, the proximity of three reacting partners accounts for the observed stereoselectivity in **252**. A ring expansion from the [3.3.0]bicyclooctane-like core can occur, followed by a simultaneous β-Et_3Si group *syn*-elimination, to give **252** in a highly regio- and stereoselective manner. The formation of **251** takes place presumably from the common intermediate **254**, which undergoes a 1,4-hydride shift.

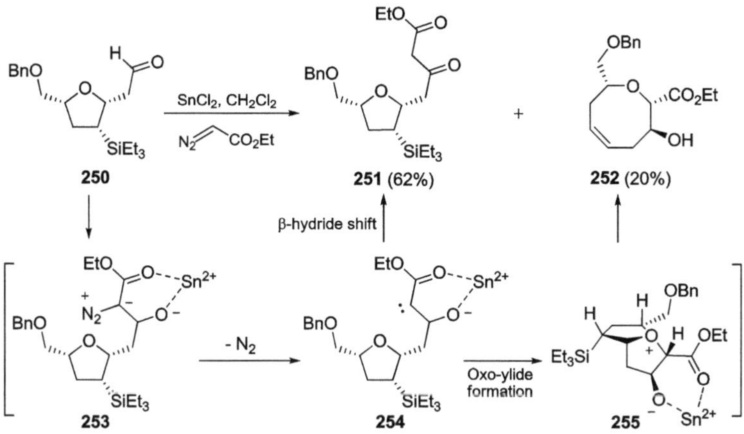

Scheme 53 Enantioselective synthesis of the Δ^4-oxocene cores via ring expansion, by Li et al.

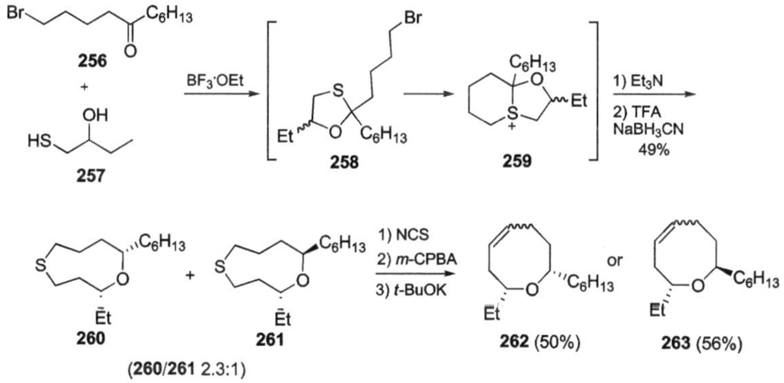

Scheme 54 Synthesis of the Δ^4-oxocene core via ring expansion by De Voss and Coster

De Voss and Coster reported the synthesis of Δ^4-oxocenes **262** and **263** through a three-step Ramberg–Bäcklund rearrangement sequence of the *cis*- and *trans*-1,5-oxathionanes **260** and **261** as the key step [168] (Scheme 54). Compounds **260** and **261** were obtained by condensation of ketone **256** with β-hydroxythiol **257** followed by the addition of a base and reduction. Conceptually related to the De Voss and Coster *O,S*-acetal fragmentation, Braddock et al. reported a transannular oxonium ion formation/fragmentation during their work towards the synthesis of the obtusallene family of natural products [169].

Recently, Njardarson et al. have reported the formation of oxocines by using a chemoselective intermolecular copper-mediated [2,3] ring expansion of a vinyl tetrahydrofuran, with ethyl diazoacetate or diazomalonate as the carbenoid precursor to generate oxonium ylides [170].

2.5 *Ring Opening and Rearrangement of Epoxides*

The opening of epoxides and rearrangements have provided efficient ways to construct medium-sized motifs during the total synthesis of many natural products. The opening of epoxides with oxygen nucleophiles, with the exception of polyepoxide-opening cascades [171], results in the oxygen of the epoxide acting as an alcohol substituent of the formed ring. Additionally, there have been several reported rearrangements of epoxide-containing substrates resulting in medium-ring oxacycles in which the epoxide oxygen becomes the ethereal oxygen in the cycle. Less commonly employed is the use of carbon nucleophiles for the opening of epoxides, as this approach requires that the ethereal bond is prepared in a separate step.

2.5.1 Epoxide Opening with Oxygen Nucleophiles

Despite that intramolecular cyclization of linear epoxy alcohols is a well-established method for the preparation of five- and six-membered rings and that this method has been limited to the preparation of larger oxepane scaffolds, Suzuki et al. have reported a method for the generation of a variety of medium-ring ethers [172]. Thus, this group extensively investigated the Lewis acid-mediated opening of epoxides and found that $Eu(fod)_3$ was suitable for the formation of eight- and nine-cyclic ethers (Scheme 55). Later, they have successfully applied this strategy to the preparation of several natural products such as (+)-(*Z*)-laureatin **266** [58] or (+)-obtusenyne **6** [173], including examples of cyclization of highly functionalized linear precursors.

Epoxide-opening processes [174] have been extensively used in the total synthesis of polycyclic ether natural product [175]. Based on the strategy of the bromonium ion-assisted epoxide ring opening [176], Braddock et al. have shown, with epoxide **267**, the feasibility of an intramolecular bromonium ion-assisted epoxide ring opening for the concurrent formation of seven-, eight-, and nine-membered ring ethers corresponding to halogenated medium-ring ethers of metabolites from *Laurencia* species (Scheme 56). This constitutes for the first time the access to a medium ring of any of these natural products by a nonenzymatic bromonium-induced cyclization process from a linear precursor [177].

Gagné et al. have developed a method to generate polyether skeletons from allene–epoxide cascade reaction promoted by Au(I) [178]. For instance, allenyl epoxide **271** provided the 9-*endo* product **272** as a single diastereoisomer under treatment with $(PhO)_3PAuCl$ and AgOTf (Scheme 57). In this example the methyl group positioning on the epoxide is controlling the ring-opening regiochemistry.

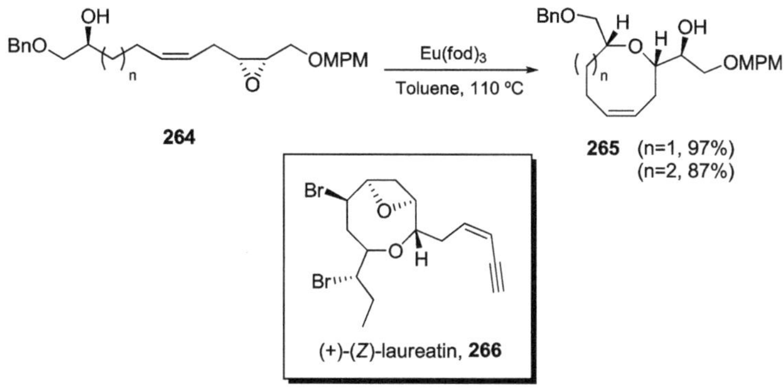

Scheme 55 Lewis acid-mediated epoxide opening in the preparation of (+)-(*Z*)-laureatin

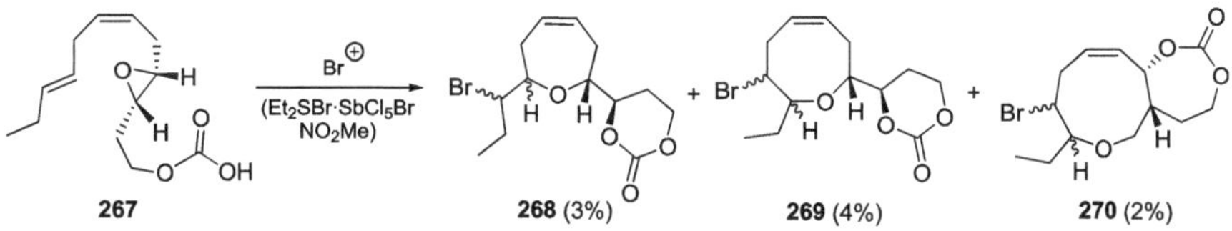

Scheme 56 Intramolecular bromonium ion-assisted epoxide ring opening for the concurrent formation of seven-, eight-, and nine-membered ring ethers, by Braddock et al.

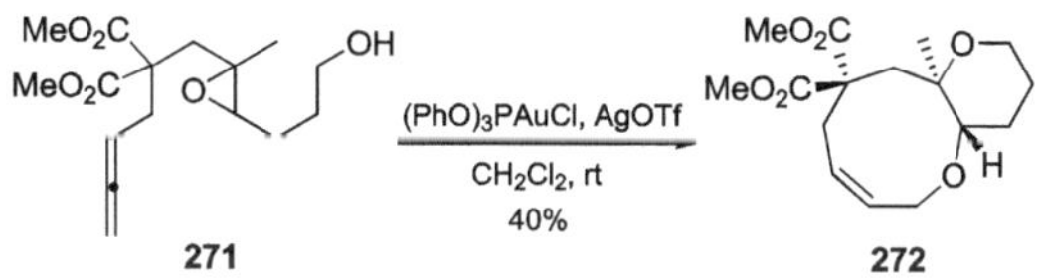

Scheme 57 Au(I)-promoted allene–epoxide cascade developed by Gagné et al.

2.5.2 Epoxide Opening with Carbon Nucleophiles

The use of carbon nucleophiles for the generation of medium-ring oxacycles is relatively rare. Chandrasekhar et al. have used this strategy for the construction of the nine-membered ring of eleutherobin analogs (Scheme 58) [179], a family of potent tubulin inhibitor molecules. Anomeric alkylation of protected D-mannose **273** with *ortho*-bromobenzyl bromide, followed by selective ketal deprotection and a two-step epoxide installation, afforded **274**. Generation of an intermediate aryl anion through a lithium–halogen exchange of **274** and then treatment with a Lewis acid afforded the oxocine product **275**.

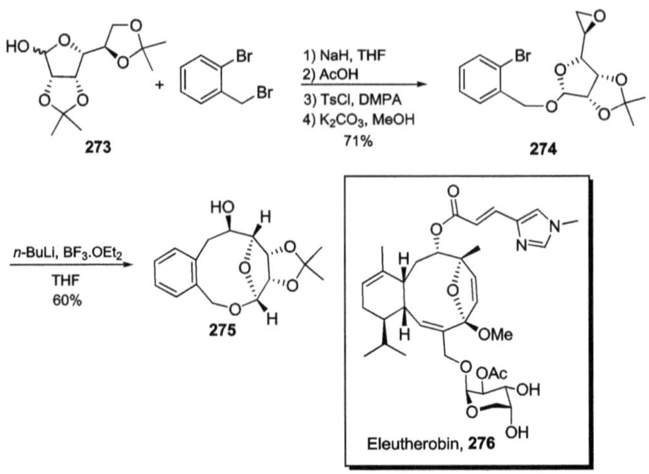

Scheme 58 Synthesis of the nine-membered ring of eleutherobin analogs by using nucleophilic carbons

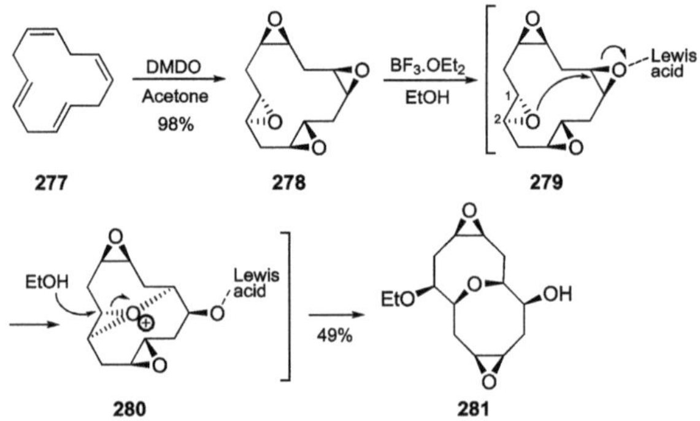

Scheme 59 Preparation of bridged bis-oxocanes via epoxide rearrangement

2.5.3 Epoxide Rearrangement

Parrain et al. disclosed the preparation of bridged bis-oxocanes from macrocyclic polyepoxides [180] on the basis of the previous studies of Simmons et al. [181] and Paquette and Vazeux [182] who reported the use of epoxide-opening cascades for the preparation of topologically nonplanar molecules via postulated [2σ+2σ+2σ] to [2σ+2σ+2σ] sigmatropic rearrangements. Thus, tetraepoxide **278**, derived from the epoxidation of all (Z)-1,4,7,10-cyclododecatetraene **277**, rearranges to form bridged bis-oxocane **281** via the formation of the epoxonium ion **280** by transannular nucleophilic attack by the C1–C2 epoxide (Scheme 59).

Martínez et al. reported the ring expansion of camphor derivatives to eight-membered ring ethers [183]. As shown in Scheme 60, the diepoxide precursor **282**

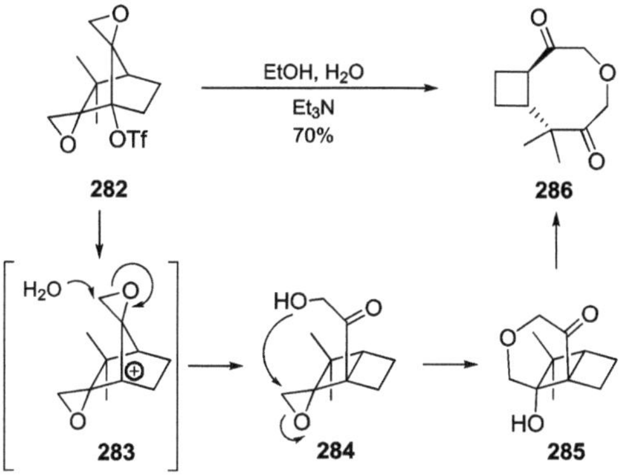

Scheme 60 Ring expansion of camphor to form eight-membered ring ethers

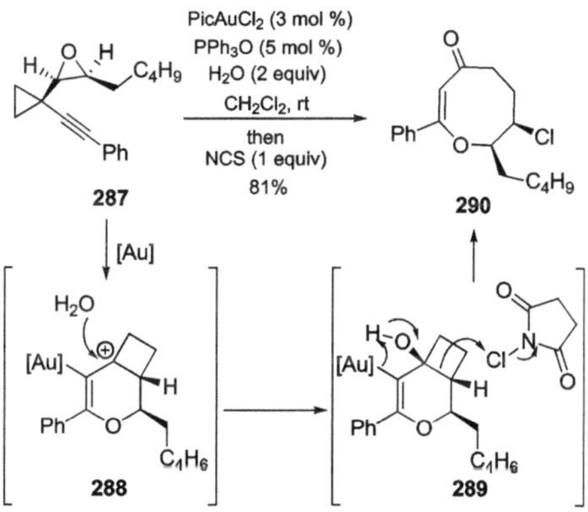

Scheme 61 Formation of eight-membered cyclic ethers via a [Au]-catalyzed isomerization of an epoxide

was heated at reflux in an aqueous ethanol buffered with triethylamine to provide the cyclobutane-fused eight-membered ring ether **286** via the ring expansion of tricyclic intermediate **285**. Compound **285** is formed by an initial triflate solvolysis of **282** that generates bridgehead cation **283**, which can then undergo an epoxide-based pinacol-type rearrangement to yield α-hydroxy ketone **284**. A subsequent 6-*exo-tet* cyclization epoxide opening affords **285** which upon undergoing a retro-aldol led to the observed fused bicyclic compound **286**.

Taking advantage of the selective activation of π-bonds by gold complexes, Liu and Liao reported the formation of an eight-membered ring ether **290** via a [Au]-catalyzed isomerization of cyclopropane epoxide **287** [184] (Scheme 61). Thus, **290** is proposed to arise from an initial [Au]-mediated ring expansion/isomerization of *cis*-1-oxiranyl-1-alkynylcyclopropane **287** to the hypothetical cation **288**. Trapping of cation **288** with water gives cyclobutanol **289**, which upon treatment with NCS furnishes the eight-membered heterocycle **290**.

3 Conclusion and Outlook

The review presented herein summarizes and updates some recent publications about the formation of eight- to ten-membered cyclic ethers. The synthesis of these systems represents a significant challenge, because both entropic and enthalpic barriers hamper cyclization strategies. Therefore, a variety of strategies and approaches have been developed in the last years to overcome these difficulties and have been successfully applied to the total synthesis of important natural products. Due to the large number of natural products containing medium-sized cyclic ether motifs, more methods are likely to be developed to access these structures in the future.

Acknowledgment ES thanks MICINN for a grant (CTQ2009-10478).

References

1. Faulkner DJ (2002) Nat Prod Rep 19: 1–48 and following reviews in the same series
2. Mandolini L (1987) Adv Phys Org Chem 22:1–111
3. Irie T, Suzuki M, Masamune T (1965) Tetrahedron Lett: 1091–1099
4. Chattopadhyay SK, Karmakar S, Biswas T, Majumdar KC, Rahaman H, Roy B (2007) Tetrahedron 63:3919–3952
5. Yet L (2000) Chem Rev 100:2963–3008
6. Raveto JL, Regueiro A, Marine JD (1992) Tetrahedron Lett 33:3389–3392
7. Roxburgh CJ (1993) Tetrahedron 49:10749
8. Snyder NL, Haines HM, Peczuh MW (2006) Tetrahedron 62:9301–9320
9. Elliott MC, Williams EJ (2001) J Chem Soc Perkin 1: 2303–2340
10. Elliott MC (2002) J Chem Soc Perkin 1: 2301–2323
11. Nakamura I, Yamamoto Y (2004) Chem Rev 104:2127–2198
12. Inoue M (2005) Chem Rev 105:4379–4405
13. Nakata T (2005) Chem Rev 105:4314–4347
14. Sasaki M (2006) Recent advances in total synthesis of marine polycyclic ethers. In: Kiyota H (ed) Topics in heterocyclic chemistry, vol 5. Springer, Heidelberg, pp 149–178
15. Fujiwara K (2006) Total synthesis of medium-ring ethers from laurencia red algae. In: Kiyota H (ed) Topics in heterocyclic chemistry, vol 5. Springer, Heidelberg, pp 97–148
16. Kleinke AS, Webb D, Jamison TF (2012) Tetrahedron 68:6999–7018
17. Hoberg JO (1998) Tetrahedron 54:12631–12670
18. Shiina I (2007) Chem Rev 107:239–273
19. Parenty A, Moreau X, Niel G, Campagne JM (2013) Chem Rev. doi:10.1021/cr300129n
20. Maier ME (2000) Angew Chem Int Ed 39:2073–2077
21. Grubbs RH (2010) Metathesis in natural product synthesis. Wiley-VCH, Weinheim
22. Schrock RR (2009) Chem Rev 109:3211–3226
23. Vougioukalakis GC, Grubbs RH (2010) Chem Rev 110:1746–1787
24. Samojiowicz C, Bieniek GK, Grela K (2009) Chem Rev 109:3708–3742
25. Monsaert S, Vial AL, Drozdzak R, Der Voort PV, Verpoort F (2009) Chem Soc Rev 38:3360–3372
26. Hoveyda AH, Zhugralin AR (2007) Nature 450:243–251

27. Drozdzak B, Allaert R, Ledoux N, Dragutan I, Dragutan V, Verpoort F (2005) Coord Chem Rev 249:3055–3074
28. Fürstner A, Langemann K (1997) J Am Chem Soc 119:9130–9136
29. Watanabe K, Suzuki M, Murata M, Oishi T (2005) Tetrahedron Lett 46:3991–3995
30. Fujiwara K, Goto A, Sato D, Kawai H, Suzuki T (2005) Tetrahedron Lett 46:3465–3468
31. Groaz E, Banti D, North M (2007) Eur J Org Chem: 3727–3745
32. Kadota I, Ueno H, Sato Y, Yamamoto Y (2006) Tetrahedron Lett 47:89–92
33. Ortega N, Martín T, Martín VS (2006) Org Lett 8:871–873
34. Domon D, Fujiwara K, Murai A, Kawai H, Suzuki T (2005) Tetrahedron Lett 46:8285–8288
35. Ortega N, Martin T, Martin VS (2009) Eur J Org Chem: 554–563
36. Fujiwara K, Kawamura N, Kawai H, Suzuki T (2009) Tetrahedron Lett 50:1236–1239
37. Sato D, Fujiwara K, Kawai H, Suzuki T (2008) Tetrahedron Lett 49:1514–1517
38. Majumdar KC, Maji PK, Rahama H, Roy B (2006) Lett Org Chem 3:845–847
39. Krauss J, Bracher F, Unterreitmeier D (2006) Arch Pharm Pharm Med Chem 339:629–632
40. Alberico D, Lautens M (2006) Synlett: 2969–2972
41. Jafarpour F, Lautens M (2006) Org Lett 8:3601–3604
42. Rudolph A, Rackelmann N, Turcotte-Savard MO, Lautens M (2009) J Org Chem 74:289–297
43. Kotha S, Dipak MK (2012) Tetrahedron 68:397–421
44. Oguri H, Sasaki S, Oishi T, Hirama M (1999) Tetrahedron Lett 40:5405–5408
45. Van Hooft PAV, Leeuwenburgh MA, Overkleeft HS, Van der Marel GA, Van Boeckel CAA, Van Boom JH (1998) Tetrahedron Lett 39:6061–6064
46. Crimmins MT, Choy AL (1999) J Am Chem Soc 121:5653–5660
47. Edwards SD, Lewis T, Taylor RJK (1999) Tetrahedron Lett 40:4267–4270
48. Elickson KL (1983) Constituents of laurencia. In: Scheuer PJ (ed) Marine natural products, chemical and biological perspectives, vol 5. Academic, New York, NY
49. Murai A, Murase H, Matsue H, Masamune T (1977) Tetrahedron Lett 18:2507–2510
50. Kadota I, Ujehara H, Yamamoto Y (2004) Tetrahedron 60:7361–7365
51. Overman LE, Thompson AS (1988) J Am Chem Soc 110:2248–2256
52. Bratz M, Bullock WH, Overman LE, Takemoto T (1995) J Am Chem Soc 117:5958–5966
53. Burton JW, Clark JS, Derrer S, Stock TC, Bendall JG, Holmes AB (1997) J Am Chem Soc 119:7483–7498
54. Robinson RA, Clark JS, Holmes AB (1993) J Am Chem Soc 115:10400–10401
55. Fujiwara K, Yoshimoto S, Takizawa A, Souma S, Mishima H, Murai A, Kawai H, Suzuki T (2005) Tetrahedron Lett 46:6819–6822
56. Crimmins MT, Elie AT (2000) J Am Chem Soc 122:5473–5476
57. Sugimoto M, Suzuki T, Haiwara H, Hoshi T (2007) Tetrahedron Lett 48:1109–1112
58. Saioh T, Suzuki T, Sugimoto M, Hagiwara H, Hoshi T (2003) Tetrahedron Lett 44:3175–3178
59. Kim H, Lee H, Lee D, Kim S, Kim D (2007) J Am Chem Soc 129:2269–2274
60. Crimmins MT, Emmitte KA (1999) Org Lett 1:2029–2032
61. Eliel EL, Wilen SH (1994) Stereochemistry of organic compounds. Wiley, New York, NY
62. Baek S, Jo H, Kim H, Kim H, Kim S, Kim D (2005) Org Lett 7:75–77
63. Lee H, Kim H, Yoon T, Kim B, Kim S, Kim H-D, Kim D (2005) J Org Chem 70:8723–8729
64. Fujiwara K, Goto A, Sato D, Ohtaniuchi Y, Tanaka H, Murai A, Kawai H, Suzuki T (2004) Tetrahedron Lett 45:7011–7014
65. Domon Fujiwara K, Souma S, Mishima H, Murai A (2002) Synlett: 1493–1495
66. Henry KJ Jr, Greico PA, Jagoe CTJ (1992) Tetrahedron Lett 33:1817–1820
67. Ikemoto N, Schreiber SL (1992) J Am Chem Soc 114:2524–2536
68. Uenishi J, Kawahama R, Yonemitsu O, Tsuji J (1998) J Org Chem 63:8965–8975
69. Kahnberg P, Lee CW, Grubbs RH, Sterner O (2002) Tetrahedron 28:5203–5208
70. Clark JS, Kettle JG (2007) Tetrahedron Lett 38:127–130
71. Clark JS, Kettle JG (2007) Tetrahedron Lett 38:123–126

72. Clark JS, Freeman RP, Cacho M, Thomas AW, Swallow S, Wilson C (2004) Tetrahedron Lett 45:8639–8642
73. Clark JS, Trevitt GP, Boyall D, Stammen B (1998) Chem Commun: 2629–2630
74. Miller SJ, Kim S-H, Chen Z-R, Grubbs RH (1995) J Am Chem Soc 117:2108–2109
75. Lindermann RJ, Seidlecki J, O'Neil SA, Sun H (1997) J Am Chem Soc 119:6919–6920
76. Chattopadhyay SK, Maity S, Panja S (2002) Tetrahedron Lett 43:7781–7783
77. Chattopadhyay SK, Pal BK, Maity S (2003) Chem Lett 32:1190–1191
78. Chattopadhyay SK, Dey R, Biswas S (2005) Synthesis: 403–406
79. Macias FA, Molinillo JMG, Varela RM, Torres A, Fronczek FR (1994) J Org Chem 59:8261–8266
80. Stefinovic M, Snieckus V (1998) J Org Chem 63:2808–2809
81. Kishuku H, Shindo M, Shishido K (2003) Chem Commun: 350–351
82. Macias FA, Molinillo JMG, Chinchilla D, Gallindo JCG (2004) In: Macias FA, Molinillo JMG, Gallindo JCG (eds) Allelopathy: chemistry and mode of action of allelochemicals. CRC, Boca Raton, FL
83. Beesley RM, Ingold CK, Thorpe JF (1915) J Chem Soc Trans 107:1080–1106
84. Aird JI, Hulme AN, White JW (2007) Org Lett 9:631–634
85. Seo Y, Cho KW, Rho JR, Shin J, Kwon BM, Bok SH, Song JI (1996) Tetrahedron 52:10583–10596
86. Davoren JE, Martin SF (2007) J Am Chem Soc 129:510–511
87. White JD, Martin WHC, Lincoln C, Yang J (2007) Org Lett 9:3481–3483
88. Varadarajan S, Mohapatra DK, Datta A (1998) Tetrahedron Lett 39:1075–1078
89. Murata M, Legrand AM, Ishibashi Y, Yasumoto T (1989) J Am Chem Soc 111:8929–8931
90. Satake M, Murata M, Yasumoto T (1993) Tetrahedron Lett 34:1975–1978
91. Hirama M, Oishi T, Uehara H, Inoue M, Maruyama M, Oguri H, Satake M (2001) Science 294:1904–1907
92. Inoue M, Hirama M (2004) Acc Chem Res 37:961–968
93. Clark JS, Conroy J, Blake A (2007) Org Lett 9:2091–2094
94. Kadota I, Abe T, Uni M, Takamura H, Yamamoto Y (2009) Tetrahedron 65:7784–7789
95. Takamura H, Nishiuma N, Abe T, Kadota I (2011) Org Lett 13:4704–4707
96. Crimmins MT, Ellis JM, Emmitte KA, Haile PA, McDougall PJ, Parrish JD, Zuccarello JL (2009) Chem Eur J 15:9223–9234
97. Kadota I, Takamura H, Nishii H, Yamamoto Y (2005) J Am Chem Soc 127:9246–9250
98. Holt DJ, Barker WD, Jenkins PR, Panda J, Ghosh S (2000) J Org Chem 65:482–493
99. Delgado M, Martin JD (1997) Tetrahedron Lett 38:6299–6300
100. Crimmins MT, Emmitte KA, Choy AL (2002) Tetrahedron 58:1817–1834
101. Crimmins MT, Powell MT (2003) J Am Chem Soc 125:7592–7595
102. Kaliappan KP, Kumar N (2003) Tetrahedron Lett 44:379–381
103. Pain C, Celanire S, Guillaumet G, Joseph B (2003) Synlett: 2089–2091
104. Niwa H, Wakamatsu K, Yamada K (1989) Tetrahedron Lett 30:4543–4546
105. Takemoto Y, Baba Y, Saha G, Nakao S, Iwata C, Tanaka T, Ibuka T (2000) Tetrahedron Lett 41:3653–3656
106. Takahashi T, Watanabe H, Kitahara T (2002) Heterocycles 58:99–104
107. Zhang W, Guo YW, Krohn K (2006) Chem Eur J 12:5122–5127
108. Sreedhar E, Venkanna A, Chandramouli N, Babu KS, Rao JM (2011) Eur J Org Chem: 1078–1083
109. Sreedhar E, Kumar RSC, Reddy GV, Robinson A, Babu KS, Rao JM, Srinivas PV (2009) Tetrahedron Asymmetry 20:440–448
110. Fernandes RA, Kattanguru P (2011) Tetrahedron Lett 52:1788–1790
111. Fernandes RA, Kattanguru P (2011) Tetrahedron Asymmetry 22:1930–1935
112. Wetzel I, Krauss J, Bracher F (2012) Lett Org Chem 9:169–174
113. Ramirez-Fernandez J, Collado IG, Hernandez-Galan R (2008) Synlett 3:339–342
114. Ghosh G, Hossain MF, Malik CK, Maity S (2010) Tetrahedron 66:9159–9164

115. Sato K, Sasaki M (2005) Org Lett 7:2441–2444
116. Clark JS, Kimber MC, Robertson J, McErlean CS, Wilson C (2005) Angew Chem Int Ed 44:6157–6162
117. Crimmins MT, McDougall PJ, Emmitte KA (2005) Org Lett 7:4033–4036
118. Fürstner A, Müller T (1997) Synlett: 1010–1012
119. Gerlach K, Quitschalle M, Kalesse M (1999) Tetrahedron Lett 40:3553–3556
120. Murga J, Falomir E, García-Fortanet J, Carda M, Marco JA (2002) Org Lett 4:3447–3449
121. Lee CW, Grubbs RH (2000) Org Lett 2:2145–2147
122. Fürstner A, Radkowski K, Wirtz C, Goddard R, Lehmann CW, Mynott R (2002) J Am Chem Soc 124:7061–7069
123. Nakamura T, Tsuboi K, Oshida M, Nomura T, Nakazaki A, Kobayashi S (2009) Tetrahedron Lett 50:2835–2839
124. Wang EC, Lin YL, Chen HM, Li SR, Kabuto C, Takeuchi Y (2006) Heterocycles 68:125–136
125. Fürstner A, Schlede M (2002) Adv Synth Catal 344:657–665
126. Izuchi Y, Koshino H, Hongo Y, Kanomata N, Takahashi S (2011) Org Lett 13:3360–3363
127. Majumdar KC, Ray K, Debnath P, Maji PK, Kundu N (2008) Tetrahedron Lett 49:5597–5600
128. Sibi MP, Patil K, Rheault TR (2004) Eur J Org Chem: 372–384
129. Mandal SK, Roy SC (2007) Tetrahedron 63:11341–11348
130. Cardenas DJ (2003) Angew Chem Int Ed 42:384–387
131. Söderberg BC, Hubbard JW, Rector SR, O'Neil SN (2005) Tetrahedron 61:3637–3649
132. Majumdar KC, Chattopadhyay B, Ray K (2007) Tetrahedron Lett 48:7633–7636
133. Chattopadhyay SK, Neogi K, Singha SK, Dey R (2008) Synlett: 1137–1140
134. Coulter MM, Dornan PK, Dong VM (2009) J Am Chem Soc 131:6932–6933
135. Wittstein K, Kumar K, Waldmann H (2011) Angew Chem Int Ed 50:9076–9080
136. Kim H, Lee H, Kim J, Kim S, Kim D (2006) J Am Chem Soc 128:15851–15855
137. Ellis JM, Crimmins MT (2008) Chem Rev 108:5278–5298
138. Sawada Y, Sasaki M, Takeda K (2004) Org Lett 6:2277–2279
139. Sasaki M, Hashimoto A, Tanaka K, Kawahata M, Yamaguchi K, Takeda K (2008) Org Lett 10:1803–1806
140. Sasaki M, Oyamada K, Takeda K (2010) J Org Chem 75:3941–3943
141. Shing TKM, Zhong YL (2006) Synlett: 1205–1208
142. Zhang L, Wang Y, Buckingham C, Herndon JW (2005) Org Lett 7:1665–1667
143. Chen Y, Ye S, Jiao L, Liang Y, Sinha-Mahapatra DK, Herndon JW, Yu ZX (2007) J Am Chem Soc 129:10773–10784
144. Mak SY, Curtis NR, Payne AN, Congreve MS, Wildsmith AJ, Francis CL, Davies JE, Pascu SI, Burton JW, Holmes AB (2008) Chem Eur J 14:2867–2885
145. Gilmour R, Prior TJ, Burton JW, Holmes AB (2007) Chem Commun: 3954–3956
146. Overman LE (1992) Acc Chem Res 25:352–359
147. Nicolaou KC, Hwang CK, Duggan ME, Nugiel DA, Abe Y, Bal Reddy K, DeFrees SA, Reddy DR, Awartani RA, Conley SR, Rutjes FPJT, Theodorakis EA (1995) J Am Chem Soc 117:10227–10238
148. Tripathi D, Pandey KS, Kumar P (2009) Tetrahedron 65:2226–2231
149. Carreño MC, Des Mazery R, Urbano A, Colobert F, Solladié G (2005) Org Lett 7:2039–2042
150. Nicolaou KC, Prasad CVC, Hwang CK, Duggan ME, Veale CA (1989) J Am Chem Soc 111:5321–5330
151. Nicolaou KC, Gunzner JL, Shi GQ, Agrios KA, Gärtner P, Yang Z (1999) Chem Eur J 5:646–658
152. Nicolaou KC (1996) Angew Chem Int Ed Engl 35:589–607
153. Matsuo G, Kawamura K, Hori N, Matsukura H, Nakata T (2004) J Am Chem Soc 126:14374–14376
154. Ko C, Hsung RP, Al-Rashid ZF, Feltenberger JB, Lu T, Yang JH, Wei Y, Zificsak CA (2007) Org Lett 9:4459–4462
155. Miyakoshi N, Ohgaki Y, Masui K, Mukai C (2007) Heterocycles 74:185–189

156. Mukai C, Ohta M, Yamashita H, Kitagaki S (2004) J Org Chem 69:6867–6873
157. Mukai C, Yamashita H, Ichiryu T, Hanaoka M (2000) Tetrahedron 56:2203–2209
158. Díaz DD, Betancort JM, Martín VS (2007) Synlett: 343–359
159. Alcaide B, Almendros P, Carrascosa R, Martínez del Campo T (2009) Chem Eur 15:2496–2499
160. Boeckman RK, Zhang J, Reeder MR (2002) Org Lett 4:3891–3894
161. Hamura T, Kawano N, Matsumoto T, Suzuki K (2006) Chem Lett 35:730–731
162. Lecornué F, Paugan R, Ollivier J (2005) Eur J Org Chem: 2589–2598
163. Ghosh S, Tuhina K, Bhowmik DR, Venkateswaran RV (2007) Tetrahedron 63:644–651
164. Sakamoto Y, Tamegai K, Nakata T (2002) Org Lett 4:675–678
165. Snyder SA, Treitler DS, Brucks AP, Sattler W (2011) J Am Chem Soc 133:15898–15901
166. Snyder SA, Brucks AP, Treitler DS, Moga I (2012) J Am Chem Soc 134:17714–17721
167. Li J, Suh JM, Chin E (2010) Org Lett 12:4712–4715
168. Coster MJ, De Voss J (2002) J Org Lett 4:3047–3050
169. Braddock DC, Millan DS, Pérez-Fuertes Y, Pouwer RH, Sheppard RN, Solanki S, White AJ (2009) J Org Chem 74:1835–1841
170. Mack DJ, Batory LA, Njardarson JT (2012) Org Lett 14:378–381
171. Vilotijevic I, Jamison TF (2009) Angew Chem Int Ed 48:5250–5281
172. Saitoh T, Suzuki T, Onodera N, Sekiguchi H, Hagiwara H, Hoshi T (2003) Tetrahedron Lett 44:2709–2712
173. Uemura T, Suzuki T, Onodera N, Hagiwara H, Hoshi T (2007) Tetrahedron Lett 48:715–719
174. McDonald FE, Tong R, Valentine JC, Bravo F (2007) Pure Appl Chem 79:281–291
175. Nicolaou KC, Frederick MO, Aversa RJ (2008) Angew Chem Int Ed 47:7182–7225
176. Davies SG, Polywka MEC, Thomas SE (1985) Tetrahedron Lett 26:1461–1464
177. Bonney KJ, Braddock DC (2012) J Org Chem 77:9574–9584
178. Tarselli MA, Zuccarello JL, Lee SJ, Gagné MR (2009) Org Lett 11:3490–3492
179. Chandrasekhar S, Jagadeshwar V, Narsihmulu C, Sarangapani M, Krishna DR, Vidyasagar J, Vijay D, Sastry GN (2004) Bioorg Med Chem Lett 14:3687–3689
180. Guiard S, Giorgi M, Santelli M, Parrain JL (2003) J Org Chem 68:3319–3322
181. Benner SA, Maggio JE, Simmons HE III (1981) J Am Chem Soc 103:1581–1582
182. Paquette LA, Vazeux M (1981) Tetrahedron Lett 22:291–294
183. Martínez AG, Vilar ET, Fraile AG, de la Moya Cerero S, Morillo CD (2007) Tetrahedron Lett 48:5185–5188
184. Liao HH, Liu RS (2011) Chem Commun: 1339–1341

Synthesis of 12- to 16-Membered-Ring Lactones

Martin Cordes and Markus Kalesse

Contents

M. Cordes • M. Kalesse (✉)
Institute of Organic Chemistry, Gottfried Wilhelm Leibniz University of Hannover,
Schneiderberg 1b, 30167 Hannover, Germany
e-mail: markus.kalesse@oci.uni-hannover.de

J. Cossy (ed.), *Synthesis of Saturated Oxygenated Heterocycles II*, Topics in Heterocyclic Chemistry 36, DOI 10.1007/978-3-642-41470-1_3,

Abstract A wide variety of natural products exhibit a macrocyclic lactone motif with a large spectrum of biological properties ranging from perfumery, over phytotoxicity, pheromone, or insecticide activity to medicinal (antibiotic, cytotoxic, antiangiogenesis) properties. These molecules feature a wide range of cyclic structures up to 60-membered rings. For this reason, the devise of new macrolactonization methods is always of general interest. The objective of this chapter is not only to present an overview of the macrolactonization of seco-acids in the total synthesis of natural products but to present other new effective procedures for the ring closure of 12- to 16-membered ring lactones.

Keywords 12- to 16-membered • Macrolactones • Lactones • Macrolactonization • Metathesis • Natural products • Total synthesis

1 Introduction

Macrolactones are frequently observed structural motifs in natural products in general and polyketides in particular. Often the macrocyclic structure is one prerequisite of their biological activity and it is proposed that the so-obtained conformational restriction might be a general strategy for pre-orientation and ultimately optimal binding of ligands to their biological target. Additionally, the so-obtained conformational restriction might add to the chemical stability of natural products due to unfavorable stereoelectronic effects that disfavor retro-aldol processes and other transformations that could potentially lead to decomposition. In that respect, macrocyclization strategies are pivotal elements for natural products syntheses. For polyketides the macrocyclization takes place at the very end of their biosynthesis and has therefore to tolerate a variety of functional groups and protecting groups. That leads in turn to different macrocyclization strategies beyond obvious lactonizations, such as RCM, RCAM, intramolecular cross-coupling, Nozaki–Hiyama–Kishi, Wittig, and HWE methods. This review will cover different strategies and comment on advantages as well as disadvantages for natural products syntheses, providing a starting point for the evaluation of synthetic routes.

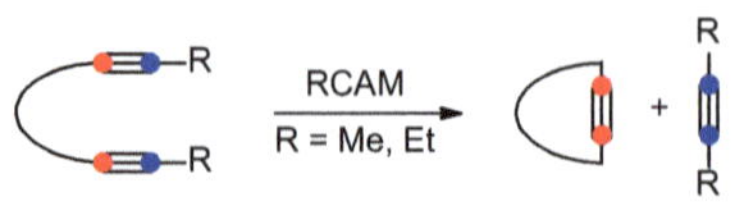

Scheme 1 Ring-closing alkyne metathesis (RCAM)

2 Ring-Closing Alkyne Metathesis

2.1 Background (RCAM)

The ring-closing alkyne metathesis (RCAM) is the reaction of an acyclic diyne affording a cyclic and an acyclic alkyne as the products (Scheme 1). Initial studies in this field were described in 1968 by Bailey [1] and coworkers utilizing a heterogeneous mixture of tungsten and silicon oxides at very high temperatures (up to 450 °C) to perform metal-catalyzed alkyne cross-metathesis. However, the first homogeneous catalytic alkyne cross-metathesis was reported not until 1974 when Mortreux and Blanchard [2] discovered that $Mo(CO)_6$ and various phenol additives such as resorcinol metathesize diphenylacetylenes at 160 °C. Even though the active form of Mortreux's instant catalyst **1** (Fig. 1) remains unknown, and in spite of harsh conditions, long reaction times, and limited functional group tolerance, this system can be ideal in cases where a substrate is robust [3], as **1** is inexpensive, air stable, and user-friendly.

On the other hand, the first well-defined alkyne metathesis catalyst was the tungsten alkylidyne complex **2** developed by Schrock in the early 1980s [4, 5]. Commercially available catalyst **2** can metathesize a broad variety of functionalized alkynes under milder conditions, typically at ambient temperature up to 90 °C [6, 7]; however, thioethers, amines, or crown ether segments were not tolerated [8]. Subsequent ligand tuning, as represented in imidazolin-2-iminato tungsten alkylidyne complex **7**, allowed the activity to be improved. Catalyst **7** represents a highly active alkyne metathesis catalyst even at room temperature [9].

A quantum leap in the development of a catalyst system with enhanced functional group tolerance was achieved in 1999 by the Fürstner group using Cummins' trisamidomolybdenum complex **4** [10–12] in RCAM. Fürstner et al. realized that **4**, primordially designed for the stoichiometric cleavage of N_2, itself does not effect any metathesis, but upon dissolving complex **4** in CH_2Cl_2 (25 equivalents per mol of **4**), the resulting trisamidomolybdenum (IV) chloride **5** (Fig. 1) efficiently catalyzes a metathetic coupling of different aliphatic as well as aromatic alkynes [8, 13]. Thus, in the presence of CH_2Cl_2, **4** reacts to form a mixture of molybdenum chloride **5** and terminal metal alkylidyne **9** (Scheme 2). Interestingly, **5** is the catalytically active component, whereas the metal alkylidyne **9** was suggested to be only a turnover reagent in one cycle of RCAM.

In 2003, Moore et al. further refined this process by introducing a reductive-recycle strategy [14, 15] (Scheme 2). The use of *gem*-dihalides such as 1,1-dichloropropane to activate precatalyst **4** results in the formation of

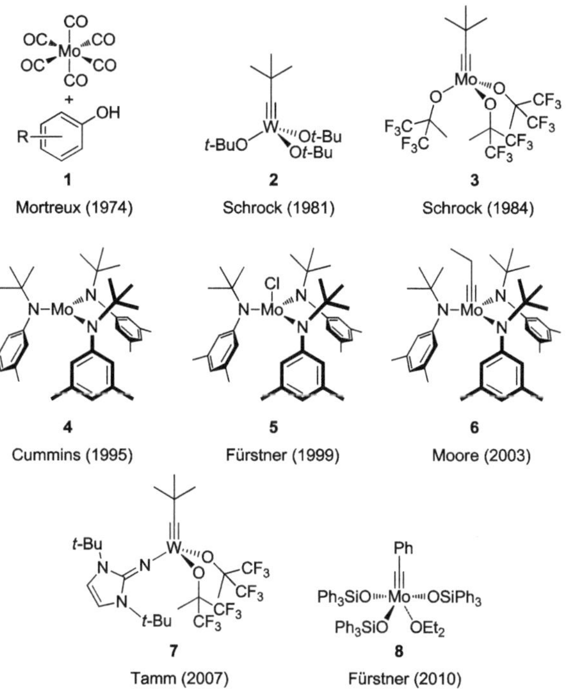

Fig. 1 Key metal complexes employed in RCAM

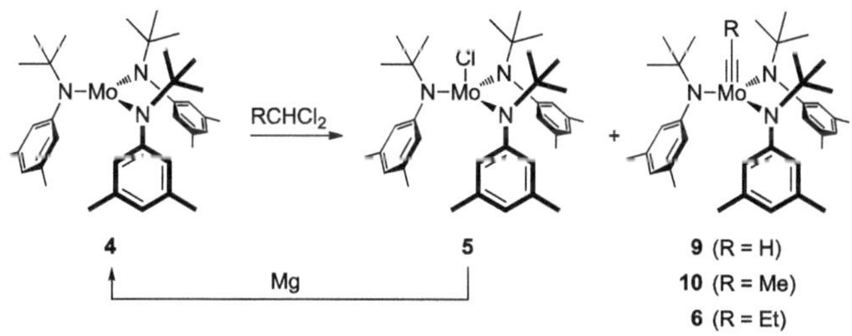

Scheme 2 Activation of molybdenum trisamido complex **4** to form metathesis-active complexes

non-terminal molybdenum alkylidyne **6** and the molybdenum chloride complex **5**, with the latter being recycled by reduction with magnesium to regenerate the parent trisamido species **4**. Furthermore, this strategy provides easy access to catalytically relevant Schrock molybdenum alkylidynes of type **3** (Fig. 1) upon alcoholysis of **6** or **10** with the appropriate alcohols and therefore creates a link to the work of Schrock et al. [16, 17] and Cummins et al. [18] on high-valent alkylidyne complexes of early transition metals. Interestingly, the terminating group has an effect on the catalytic efficacy of these systems, with R = Et (catalyst **6**) superior to R = Me (catalyst **10**).

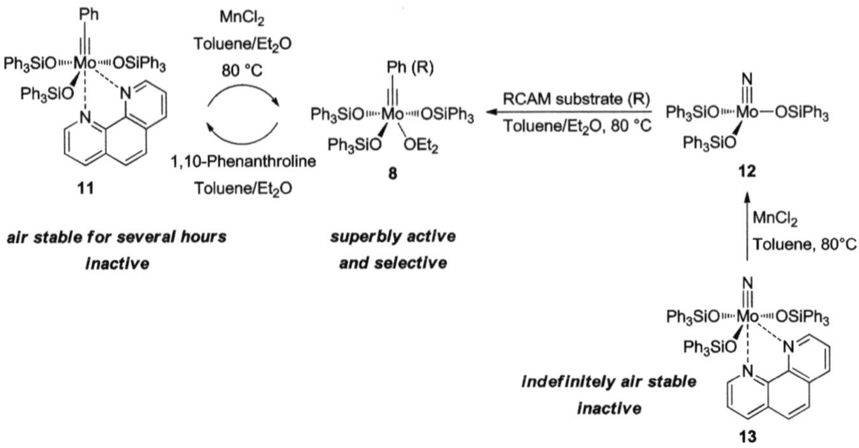

Scheme 3 Third generation RCAM catalyst **8**

A serious obstacle in using catalysts **2–7** in RCAM is the fact that all of these structures suffer from high sensitivity toward oxygen, moisture, and, in the case of complex **4**, even molecular nitrogen. Therefore, the next generation of alkyne metathesis catalysts should ideally be applicable to complex and polysubstituted targets while being inexpensive, easy to make, and air stable. This truly is a noble ambition and Fürstner et al. probably obtained, in 2010, evolving air stable (several hours) metal complex **11** from which superbly active and highly selective Schrock alkylidyne complex **8** can be released very easily [19] (Scheme 3).

Complex **8** constitutes one of the most active catalysts known to date, but retains an outstanding tolerance for functional groups. The only weakly donating triphenylsilanolate ligands impart a well-balanced level of Lewis acidity onto the d^0-molybdenum center, which is required for high catalytic activity yet is not high enough to endanger polar substituents. At the same time, the sheer size of the Ph_3Si residues prevents more than one alkyne from binding to the metal center and hence disfavors competing polymerization pathways while likely facilitating the cycloreversion of the metallacyclobutadiene intermediates.

On the other hand, the indefinitely air stable molybdenum nitride **13** also performed remarkably well. The price to be paid for the use of this fully air stable precatalyst, however, was a larger catalyst loading (generally 10 mol%), a higher reaction temperature (usually 80 °C), and longer reaction times.

2.2 12-Membered Macrocyclic Lactones

2.2.1 Cruentaren A

The total synthesis of cruentaren A **20** is a meaningful example illustrating RCAM catalyst evolution in a very short time frame. In 2007, Maier et al. utilized

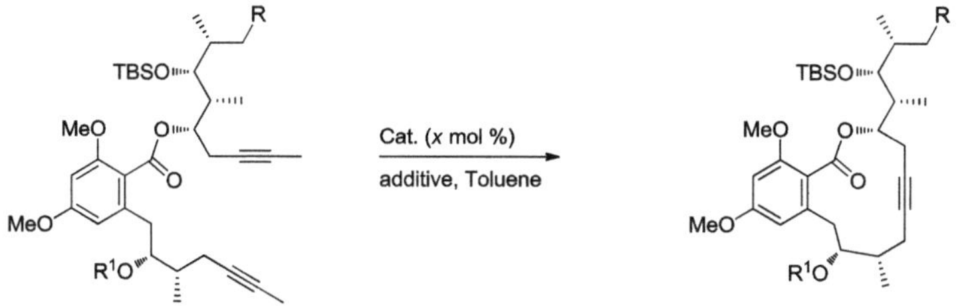

14,16,18 15,17,19

	R	R1	Cat.	x	Additive	Temp (°C)	Time (h)		Ref.
14	CH_2ODMB	TIPS	**2**	10	—	85	2	**15** (91%)	20,21
16	$(CH_2)_2ODMB$	TIPS	**2**	10	—	85	3	**17** (90%)	22
18	(*Z*)-CH=CHCH_2OTHP	TBDPS	**4**	10	CH_2Cl_2	80	1	**19** (87%)	23,24
18	(*Z*)-CH=CHCH_2OTHP	TBDPS	**12**	20	MS 4 Å	80	—	**19** (81%)	24
18	(*Z*)-CH=CHCH_2OTHP	TBDPS	**8**	2	MS 5 Å	80	—	**19** (82%)	19

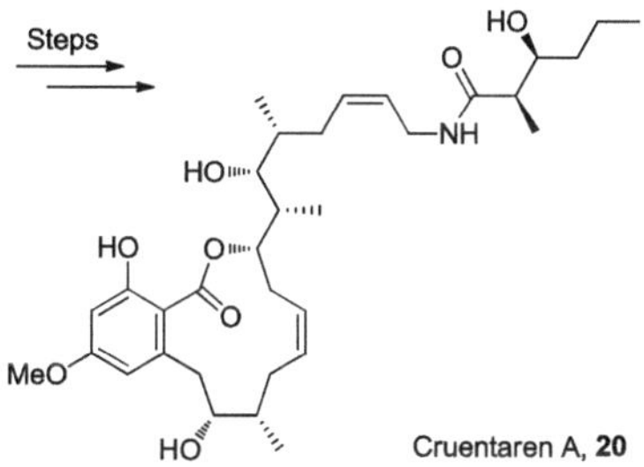

Cruentaren A, **20**

Scheme 4 Synthesis of cruentaren A, **20** by RCAM strategy

Schrock's tungsten alkylidyne complex **2** to perform RCAM with cyclization precursor **14** [20, 21] and **16** [22], whereas Fürstner et al. used Cummins' trisamido-molybdenum complex **4** for the transformation of **18–19**, due to the incompatibility of the acid-labile OTHP functionality in **18** with complex **2** [23, 24]. Further studies by Fürstner et al. in 2010 revealed superbly active catalyst **8** as ideal RCAM promoter for the generation of alkyne **19**, as **8** operates with catalyst loadings as little as 2 mol% [19]. Complex **12** [24] also performed remarkably well while needing a significant higher catalyst loading (20 mol%) (Scheme 4).

2.2.2 Lactimidomycin

Considering the substantial increase in strain energy that can be built-up by RCAM, it is noteworthy to point out that the action of catalyst **8** (5 mol%) on diyne **21** cleanly afforded the desired 12-membered enyne **22** as the sole product, which was isolated in 95 % (0.6 mmol scale) and 84 % (3.4 mmol scale) yield [25]. These cyclizations were usually performed at 80 °C for 3 h, albeit catalyst **8** per se is fully operative at ambient temperature. However, the reaction at room temperature required 24 h to go to completion [19] (Scheme 5).

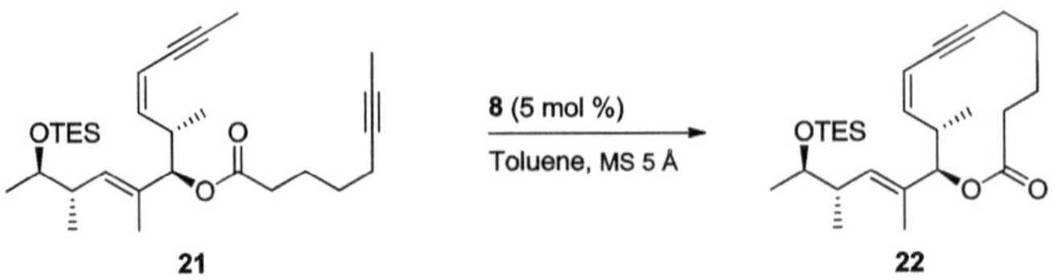

Scale (mmol)	Temp (°C)	Time (h)	Yield **22** (%)	Ref.
3.4	rt	24	84	19
3.4	80	3	84	25
0.6	80	3	95	25

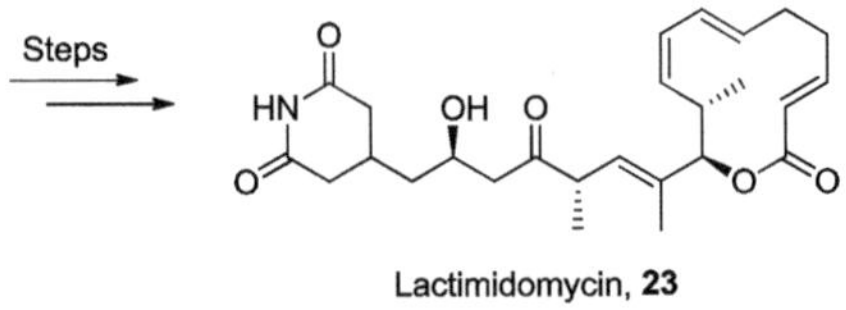

Scheme 5 Synthesis of lactimidomycin, **23** by RCAM strategy involving enyne–yne coupling

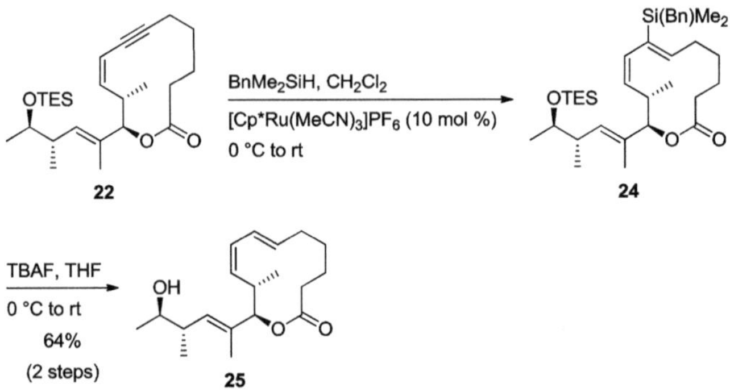

Scheme 6 Post-RCAM entry to *E*,*Z*-configured 1,3-diene **25**

The stereoselective access to (*E*)-alkene **25** was achieved by a semi-reduction method, which employed a ruthenium-catalyzed *trans*-hydrosilylation followed by a protodesilylation [25] (Scheme 6).

2.3 13-Membered Macrocyclic Lactones

2.3.1 PGE_2-1,15-lactone

The intrinsically acid- and base-labile β-hydroxy ketone moiety rendered the 13-membered prostaglandin PGE_2-1,15-lactone **28** as a probe for the applicability of the RCAM/Lindlar reduction strategy using catalyst **4**. The key transformation

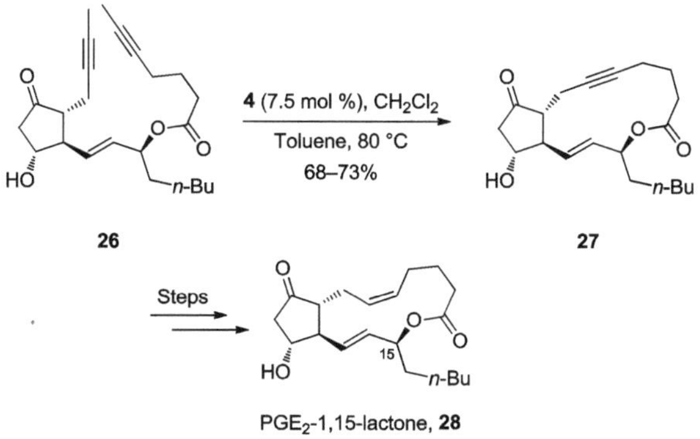

Scheme 7 Synthesis of PGE_2-1,15-lactone, **28** by RCAM strategy

using the catalyst formed in situ from **4** and CH_2Cl_2 proceeded well in toluene at 80 °C, affording the desired cycloalkyne **27** in 68–73 % yield [26] (Scheme 7).

A slightly modified catalyst in which the 3,5-dimethylphenyl group on the amido ligands of **4** were replaced by a 4-fluorophenyl residue gave an even higher yield (77 %) [27]. Standard Lindlar hydrogenation of cycloalkyne **27** followed by deprotection of the residual TBS ether with aqueous HF in acetonitrile furnished PGE_2-1,15-lactone **28**.

2.4 *14-Membered Macrocyclic Lactones*

2.4.1 *ent*-Amphidinolide V

The total synthesis of the 14-membered macrolactone *ent*-amphidinolide V **31** demonstrates once again that Fürstner's third-generation RCAM catalyst **8** is clearly superior to Cummins' trisamidomolybdenum complex **4**, as **8** operates with 2 mol% catalyst loading at ambient temperature [19], whereas **4** needs 20 mol% at 85 °C to proceed well [28] (Scheme 8). Also, the use of **8** substantially increased the yield to 81 %.

2.5 *15-Membered Macrocyclic Lactones*

2.5.1 Epilachnene

Epilachnene precursor **33** was easily accessible with either Mortreux's instant catalyst **1** or Schrock's tungsten alkylidyne complex **2** in similar yields [29] (Scheme 9). Lindlar reduction of **33** followed by deprotection of the *N*-Fmoc group with TBAF·$3H_2O$ then afforded the natural product **34**. This is in contrast to the RCM approach which gave the undesired (*E*)-isomer of **34** as the major

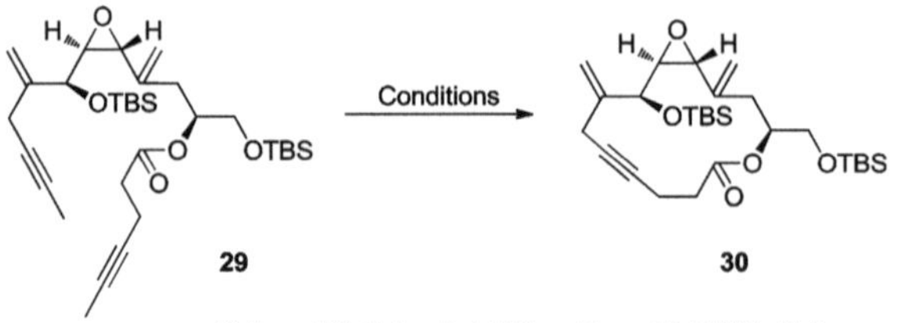

Cat.	mol %	Solvent	Additive	Temp	Yield (%)	Ref.
4	20	toluene	CH_2Cl_2	85 °C	66	28
8	2	toluene	MS 5 Å	rt	81	19

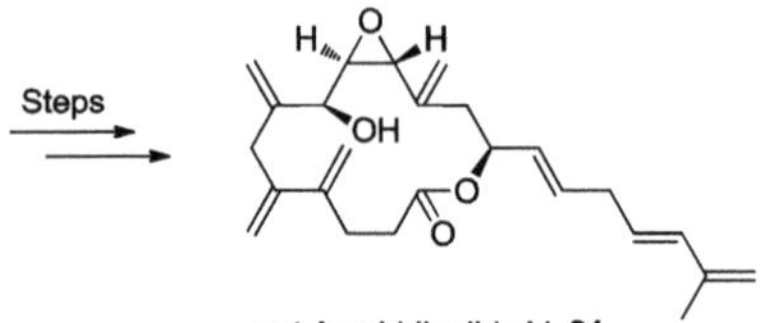

ent-Amphidinolide V, **31**

Scheme 8 RCAM-based synthesis of enantiomeric amphidinolide V, **31**

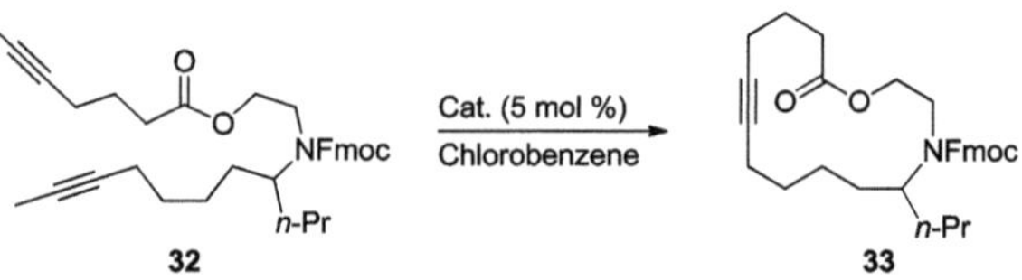

Cat.	Additive	Temp (°C)	Time (h)	Yield (%)
1 (R = 4-Cl)	4-ClC_6H_4OH (1 equiv)	132	26	67
2	–	80	1	71

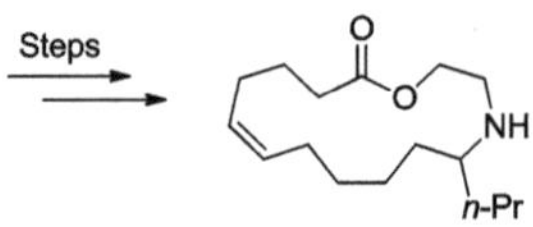

Epilachnene, **34**

Scheme 9 RCAM-based synthesis of epilachnene, **34**

constituent and underscores the basic dilemma of RCM in applications to stereochemically defined molecules [30] (cf. Sect. 3.5.1) (Scheme 16).

2.6 16-Membered Macrocyclic Lactones

2.6.1 Epothilone A

The potent biological activity of the epothilones rendered them important targets for the synthetic community and the very early total syntheses of epothilone A **39**

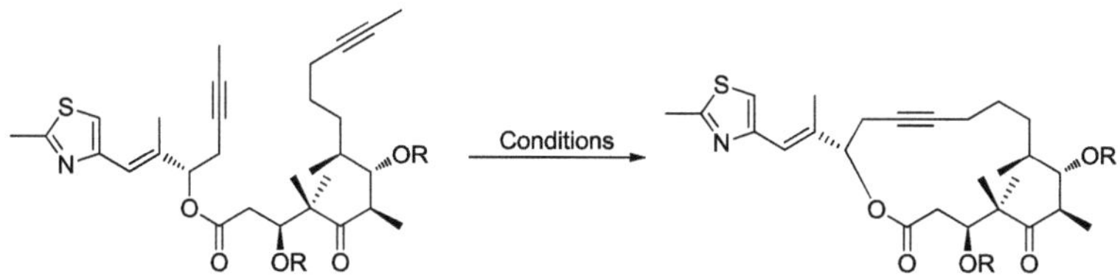

35 (R = TBS) **36** (R = TBS)

Cat.	mol %	Solvent	Additive	Temp	Time (h)	Yield (%)	Ref.
4	10	Toluene	CH_2Cl_2	80 °C	8	80	34
8	2	Toluene	MS 5 Å	rt	24	91	19

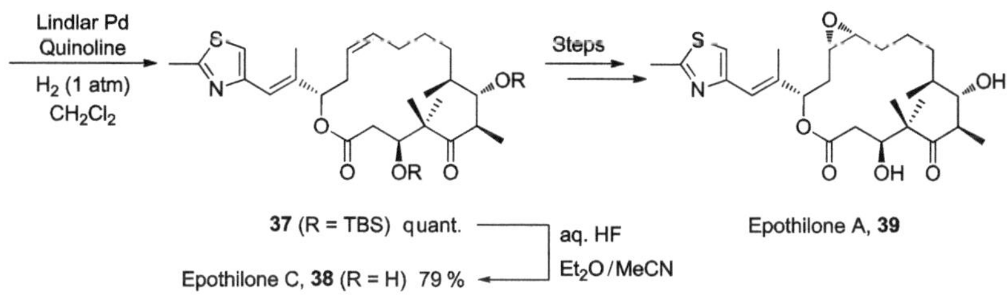

Scheme 10 RCAM-based synthesis of epothilones

were realized by using ring-closing metatheses (RCM) to form the 16-membered macrocyclic lactone [31–33]. A drawback in all these efforts was the absence of any significant selectivity for the desired (*Z*)-olefin geometry at a late stage of the synthesis.

Therefore, the epothilones serve as a valuable educational example for the potential applicability of a ring-closing alkyne metathesis/Lindlar reduction sequence to perform selective access to (*Z*)-alkene **37** (Scheme 10). In fact, diyne **35** was smoothly converted to the 16-membered cycloalkyne **36** in 80 % isolated yield on exposure to catalytic amounts of molybdenum amido complex **4** in toluene/CH_2Cl_2 at 80 °C [34]. Lindlar reduction of cycloalkyne **36** followed by cleavage of the silyl ether groups in the resulting (*Z*)-alkene **37** by aq. HF in Et_2O/MeCN delivers epothilone C **38** in 79 % yield.

Besides the proper outcome of the alkene geometry, additional aspects favor this strategy over alternative routes as (1) neither the basic *N*-atom nor the sulfur group of the thiazole ring interfere with the catalyst; (2) the labile aldol substructure, the rather electrophilic ketone, as well as the ester and silyl ether groups are fully preserved; (3) no racemization of the chiral center α to the carbonyl is encountered; and (4) the rigorous chemoselectivity of the catalyst is confirmed, which reacts smoothly with alkynes but leaves preexisting alkene moieties unaffected.

In a recent approach, Fürstner et al. used catalyst **8** as RCAM promoter for the generation of cycloalkyne **36** [19]. As previously shown for complex **4**, catalyst **8** also reacts with exquisite chemoselectivity in the cyclization event, rigorously distinguishing between the alkene and alkyne π-bonds of starting material **35**.

Nevertheless, the most important advantages of **8** were the excellent yields (91 %), the low catalyst loading (2 mol%), and the fact that these reaction can be run at ambient temperature, albeit 24 h was necessary for the RCAM transformation to go to completion (Scheme 10).

2.6.2 Latrunculin A

Due to the known incompatibility of complex **4** with *N*-unprotected amides, the Teoc derivative **40** allowed RCAM to proceed with high chemoselectivity at the triple bonds to form the highly strained 16-membered cyclic product **41** in 70 % yield [35–37]. In this context, it is worth mentioning that standard alkene metathesis catalysts do not distinguish between alkenes and alkynes, but attack both types of π systems with similar rates. In summary, this route featured the first successful implementation of a ring-closing enyne–yne metathesis reaction into a total synthesis of macrocyclic lactones (Scheme 11). *Z*-Selective hydrogenation of the triple bond in **41** with Lindlar's catalyst, in the presence of a large excess of quinoline to suppress over-reduction, yielded (*Z*)-alkene **42** in 82 %.

3 Ring-Closing Olefin Metathesis

3.1 Background (RCM)

In ring-closing olefin metatheses (RCM), formally, bond fission of two double bonds occurs and the new double bond is formed at the same time to produce a cyclic compound (Scheme 12).

Although RCM has proved to be particularly useful for the synthesis of medium-sized (5–10-membered) rings, this tool was also used for macrocycles exhibiting ring sizes of 11-membered rings or more. For five- to eight-membered rings generally thermodynamic (*Z*)-alkenes were formed, while larger rings were obtained as *E*/*Z* mixtures where the (*E*)-isomer was dominant. The stereochemical outcome of the newly formed double bond for 10- to 16-membered rings is a nontrivial problem since the factors that control the *E*/*Z* selectivity still remain unclear [38].

A large number of catalysts have been developed and employed in RCM over the past decades [39]. Figure 2 shows only those metal alkylidene complexes that have been frequently employed in RCM to 12- through 16-membered lactones, namely, Schrock catalyst **44** and the Grubbs catalysts **45**, **46**, **47** and **48**.

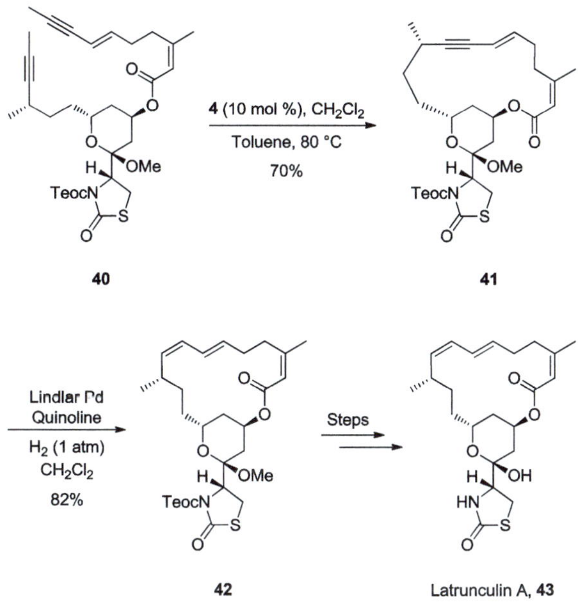

Scheme 11 RCAM/Lindlar hydrogenation approach to latrunculin A, **43**, involving enyne–yne coupling

Scheme 12 Ring-closing olefin metathesis (RCM)

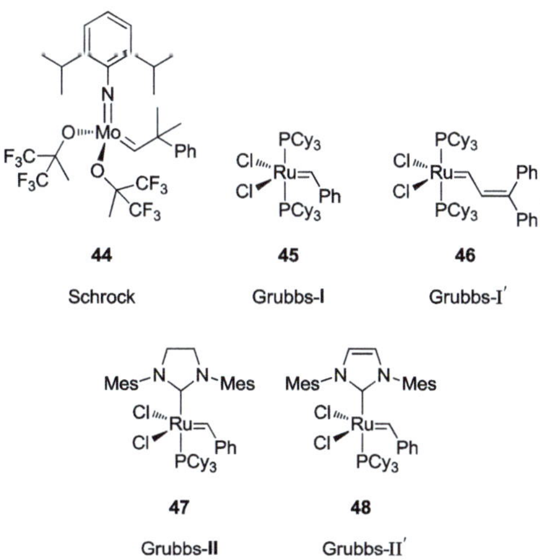

Fig. 2 Selected metal alkylidene complexes that have been employed in RCM to 12- through 16-membered macrocyclic lactones

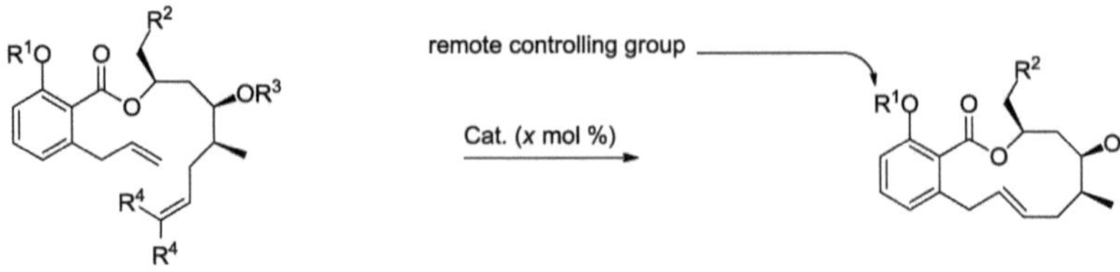

49–56 **57–64**

Entry		R¹	R²	R³	R⁴	Cat.	*x*	Solvent	T (°C)	t (h)		*E*/*Z*	Ref.
1	**49**	Me	(*E*)-CH=CHCH$_2$OPMB	TBS	H	**45**	10	CH$_2$Cl$_2$	rt	0.67	**57** (85%)	10/1	42
2	**50**	Me	CH$_2$OPMB	MOM	H	**45**	10	CH$_2$Cl$_2$	rt	1.3	**58** (93%)	9/1	43
3	**51**	MOM	CH$_2$OPMB	MOM	H	**45**	10	CH$_2$Cl$_2$	rt	1.3	**59** (93%)	9/1	43
4	**50**	Me	CH$_2$OPMB	MOM	H	**47**	10	CH$_2$Cl$_2$	rt	1.3	**58** (93%)	2/1	43
5	**51**	MOM	CH$_2$OPMB	MOM	H	**47**	10	CH$_2$Cl$_2$	rt	1.3	**59** (93%)	2/1	43
6	**52**	Me	CH$_2$OPMB	MOM	Me	**48**	5	toluene	80	1.5	**60** (93%)	2/1	41
7	**53**	MOM	CH$_2$OPMB	MOM	Me	**48**	10	toluene	80	3	**61** (91%)	2/1	41
8	**54**	TBS	CH$_2$OPMB	MOM	Me	**48**	10	toluene	80	1	**62** (91%)	1/1.5	41
9	**55**	H	(*E*)-CH=CHCO$_2$Me	H	H	**45**	—	CH$_2$Cl$_2$	—	96	**63** (—)	*Z*	40
10	**56**	H	CH$_2$OPMB	MOM	Me	**48**	10	toluene	80	20	**64** (69%)	*Z*	41

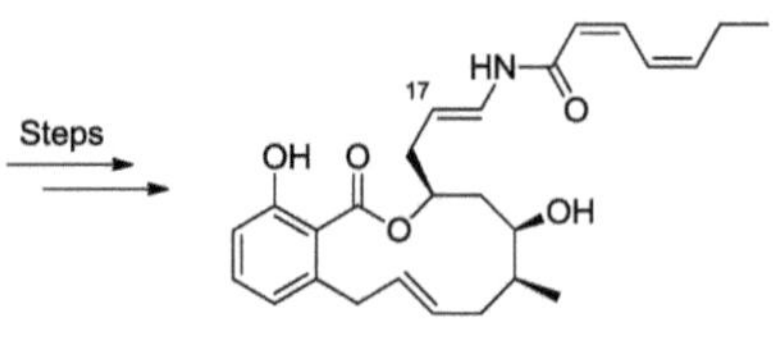

(–)-Salicylihalamide A, **65** and B, **66** (17Z-isomer)

Scheme 13 RCM optimization studies for the synthesis of salicylihalamide precursors

3.2 12-Membered Macrocyclic Lactones

3.2.1 Salicylihalamides

As aforementioned, RCM-based macrocyclizations are known to be reversible processes and usually provide *E*/*Z* mixtures, often favoring the thermodynamically more stable isomer. This issue is often the pivotal subject in order to obtain good *E*/*Z* stereoselectivities. A persuasive example of thoroughly optimization studies can be found in the synthesis of the salicylihalamides A **65** and B **66** (Scheme 13). The remote phenolic hydroxy group is of considerable interest as it favors the undesired (*Z*)-isomer (Scheme 13, entries 9 [40] and 10 [41]). If the phenolic hydroxy group is protected, especially as a methyl or MOM ether, the (*E*)-isomer is the major product [42, 43] (Scheme 13, entries 1–3). The influence of the catalyst is also crucial, as the first generation Grubbs catalyst **45** kinetically induced the formation of the desired (*E*)-isomers (Scheme 13, entries 1–3), while the second generation catalysts **47** and **48** rapidly produced a thermodynamic ratio of products indicating that an isomerization process occurred during the experiment (Scheme 13, entries 4–8) [41, 43].

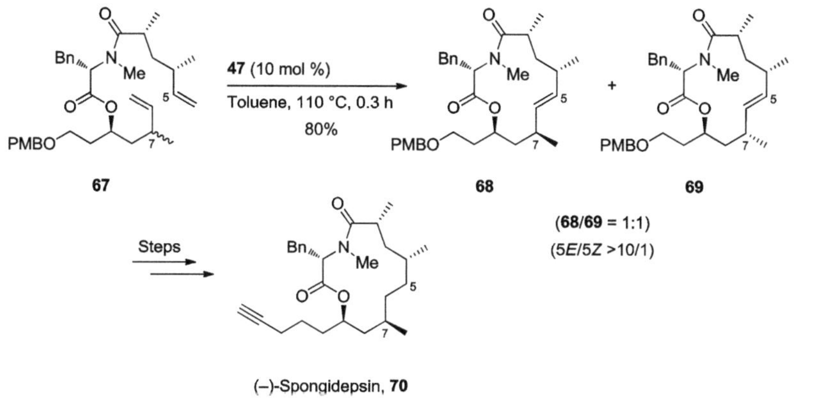

Scheme 14 RCM-based synthesis of (–)-spongidepsin, **70**

3.3 13-Membered Macrocyclic Lactones

3.3.1 (–)-Spongidepsin

Five total syntheses of the 13-membered lactone and lactam antiproliferative marine metabolite (–)-spongidepsin **70** have been reported to date [44–48], and most of them use an RCM. As a general trend, when an RCM was performed on substrates of type **67** (Scheme 14), high yields were obtained but with variable *E*/*Z* ratios. This outcome was fortunately not detrimental as the resulting double bond was finally removed by hydrogenation. Thus, a stereodivergent total synthesis strategy featuring RCM as the key step was employed and culminated in the full structural elucidation of spongidepsin relying on the preparation of all possible diastereomers of the final compound. Hence, exposure of **67** to the second generation Grubbs catalyst **47** in refluxing toluene yielded the four possible macrocycle 5*E*/*Z*,7*R*/*S* diastereomers in 80 % combined yield. The two (*E*)-alkenes, **68** and **69**, were obtained in a separable 1/1 ratio and predominated over the (*Z*)-isomers by >10/1 [44] (Scheme 14).

3.4 14-Membered Macrocyclic Lactones

3.4.1 (+)-Migrastatin

In the case of (+)-migrastatin **73**, a 14-membered macrolide, isolated from *S. platensis*, RCM was highly chemoselective and proceeded in a highly (*E*)-selective fashion [49] (Scheme 15).

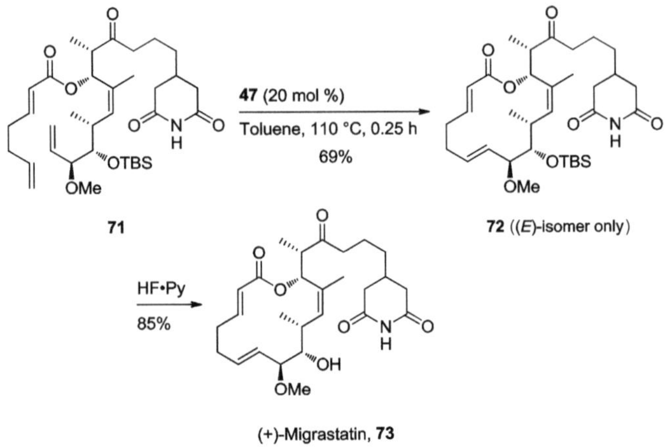

Scheme 15 (*E*)-Selective RCM-based synthesis of (+)-migrastatin, **73**

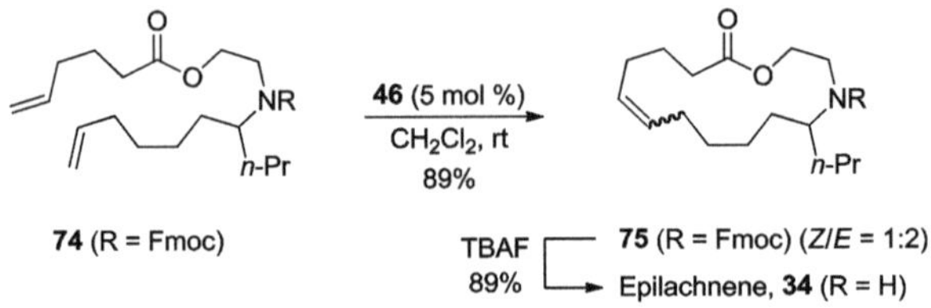

Scheme 16 RCM-based synthesis of epilachnene, **34**

3.5 15-Membered Macrocyclic Lactones

3.5.1 Epilachnene

The very salutary example of epilachnene **34** [30], a 15-membered macrolide, once again reflects the central issue of RCM that neither the control of the newly formed double bond nor the ability to its proper prediction can be guaranteed definitely (Scheme 16) (see also Sect. 2.5.1, Scheme 9).

3.6 16-Membered Macrocyclic Lactones

3.6.1 Epothilones

Three RCM-based strategies have been investigated since Danishefsky's first total synthesis of epothilone A **39** in 1996 [50], namely, the C9–C10, the C10–C11, and the C12–C13 double-bond disconnection. As can be seen from the three very early

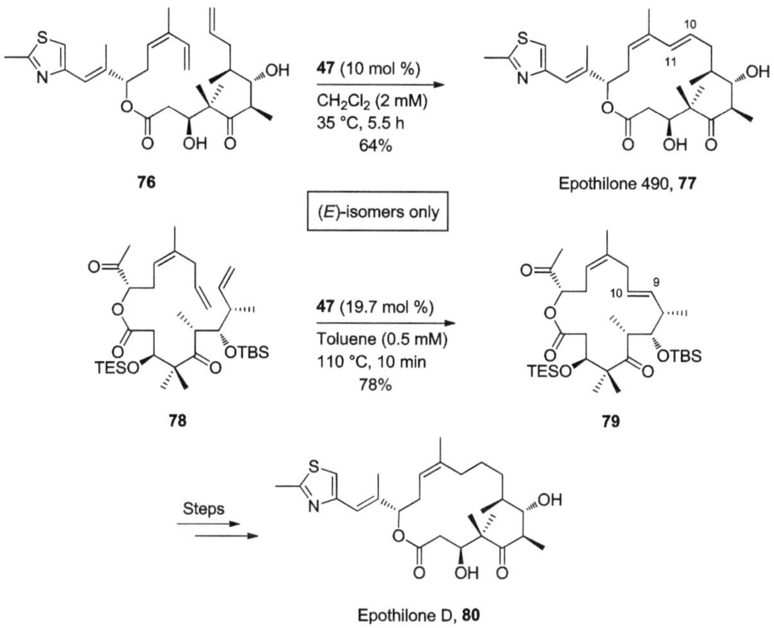

Scheme 17 Highly stereoselective RCM-based formation of (*E*)-olefins in the total synthesis of epothilone 490, **77**, and epothilone D, **80**

approaches to epothilone A **39** [31–33], the stereocontrol of the C12–C13 double bond by using RCM is difficult; indeed, none of the catalyst used initially, **44** and **45**, produced satisfactory *Z* selectivity. Since this serious problem arose only toward the very end of the rather laborious synthesis and since the isomeric alkenes could not be readily separated at this stage, it is hardly surprising that alternative double-bond disconnections were scrutinized, ensuring better control over all structural elements. However, alternative approaches involving the formation of either the C10–C11 [51] or the C9–C10 [52] double bond followed by a hydrogenation step were not possible until the discovery of more active second generation ruthenium catalysts, such as **47** (Scheme 17).

4 Alkyl Alkynyl Ethers as Ketene Source for Lactone Formation

4.1 Background

A highly innovative method for the preparation of macrocyclic lactones, namely, the thermolysis of an alkoxyacetylene followed by intramolecular trapping of the resulting ketene, has been reported by Funk et al. in the course of mechanistic investigations [53] (Scheme 18).

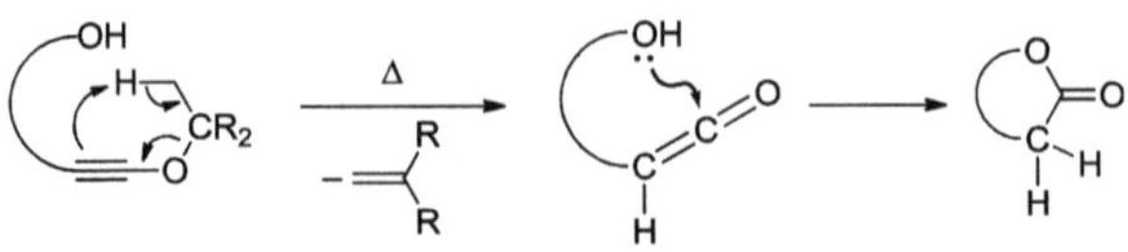

Scheme 18 Thermolysis of alkyl alkynyl ethers into ketenes and following lactonization

This technique to form macrolactones was then further improved (solvent, amine additive) but not used in the context of total synthesis of 13- and 15-membered lactones [54, 55].

4.2 16-Membered Macrocyclic Lactones

4.2.1 (+)-Acutiphycin

The first application of an alkyl alkynyl ether as a macrolactone precursor in total synthesis was reported by Jamison et al. in 2006 [56]. As shown in Scheme 19, slow addition of **81** to refluxing xylenes and Bu_3N effected a thermal retro-ene reaction to form ethylene and ketene **82** that then underwent a highly group-selective coupling with the least hindered (yet most remote) of the four hydroxyl groups present in the molecule to give the desired macrocycle **83** in excellent yield (90 %).

5 Macrolactones via Intramolecular Capture of Acylketenes

5.1 Background

Dioxolenone or β-keto ester thermolysis are known processes for obtaining β-acetylketene derivatives under relatively mild conditions (refluxing toluene) [57]. The resulting acylketene intermediate can be trapped intramolecularly by an oxygen nucleophile to give the corresponding macrolactone (Scheme 20).

Boeckman et al. were the first to use dioxinones as precursors to acylketenes in the synthesis of complex natural products featuring macrolactones, e.g., 14-membered ring, (–)-kromycin [58]. This method has found synthetic applications in the synthesis of deschlorocallipeltoside A [59] and was chosen as a key step in various syntheses of (–)-callipeltoside A **88** (14-membered) [60–63].

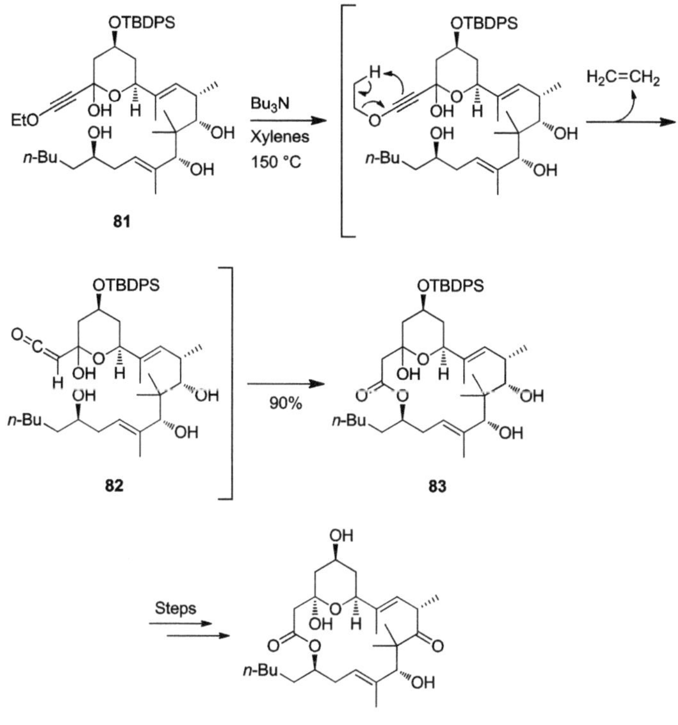

Scheme 19 Thermal retro-ene reaction of an ethyl alkynyl ether as convenient ketene source for lactone formation in the total synthesis of (+)-acutiphycin, **84**

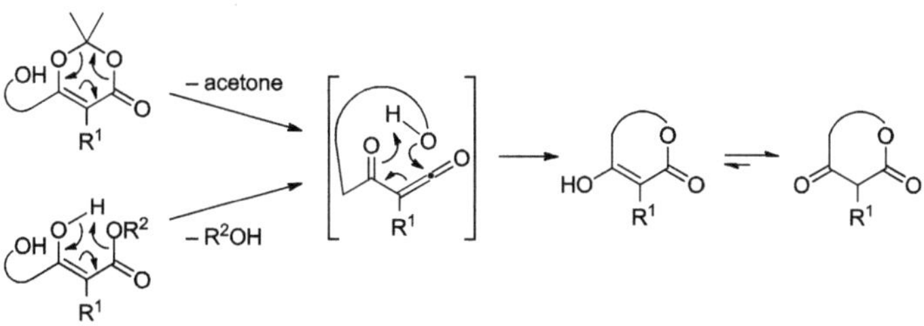

Scheme 20 Macrolactonization via intramolecular concerted addition of oxygen nucleophiles to acylketenes

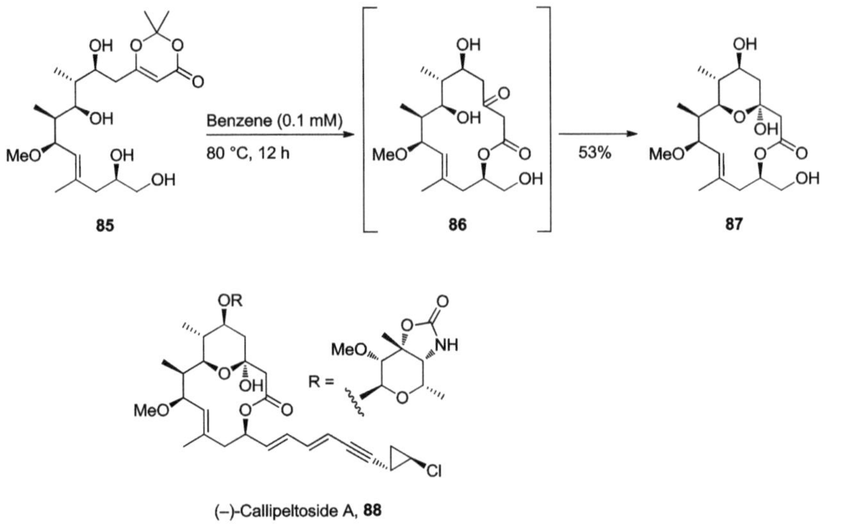

Scheme 21 Remarkable chemoselectivity in a tandem macrolactonization–hemiketalization reaction

5.2 *14-Membered Macrocyclic Lactones*

5.2.1 (–)-Callipeltoside A

The antitumor natural product (–)-callipeltoside A **88** has been a popular aim for total synthesis due to its complex architecture and promising bioactivity. The most impressive contribution to this target, concerning macrolactonization strategy, was probably made by Hoye and coworkers to accomplish both the macrolactonization and the transannular hemiketal formation as a one-pot protocol without the need of protecting groups [62] (Scheme 21).

5.2.2 Norzoanthamine

A very recent example of β-keto ester thermolysis has been shown during the synthesis of the carbocyclic core of norzoanthamine (Scheme 22). Macrocyclic lactone **90** was then further elaborated to **91** using a transannular Michael reaction cascade [64].

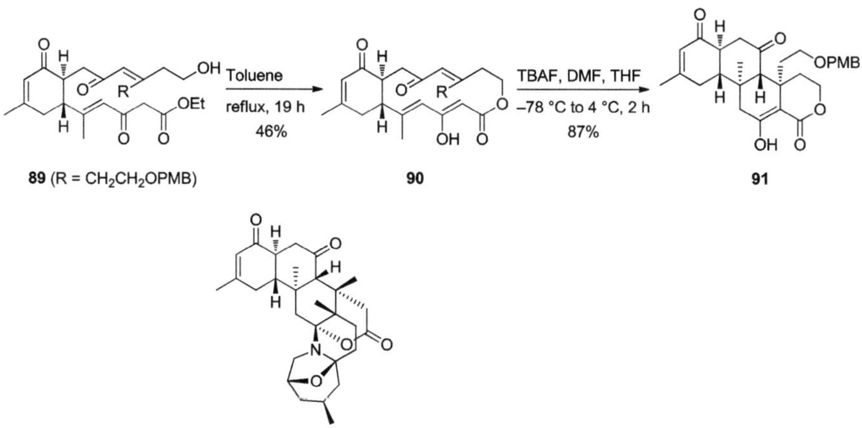

Norzoanthamine, **92**

Scheme 22 β-Keto ester thermolysis and following macrolactonization in the total synthesis of norzoanthamine, **92**

5.3 15-Membered Macrocyclic Lactones

5.3.1 (–)-Amphidinolide P

In the total synthesis of (–)-amphidinolide P **95**, an intramolecular transesterification was carried out by heating hydroxy β-keto ester **93** in refluxing toluene, which gave pure macrocycle **94** in 72 % yield certainly via a ketene intermediate [65] (Scheme 23).

5.4 16-Membered Macrocyclic Lactones

5.4.1 (–)-A26771B

In a more recent application of dioxolenone thermolysis, Blechert et al. reported a key acylketene lactonization which enables the synthesis of the 16-membered macrolide antibiotic (–)-A26771B **98**. Refluxing a dilute solution (125 μM) of **96** in *n*-heptane initiated the formation of an acylketene which was trapped by the pendant secondary alcohol to give macrolactone **97** [66] (Scheme 24). This advanced intermediate could then be elaborated into (–)-A26771B **98** over the course of six steps in high yield.

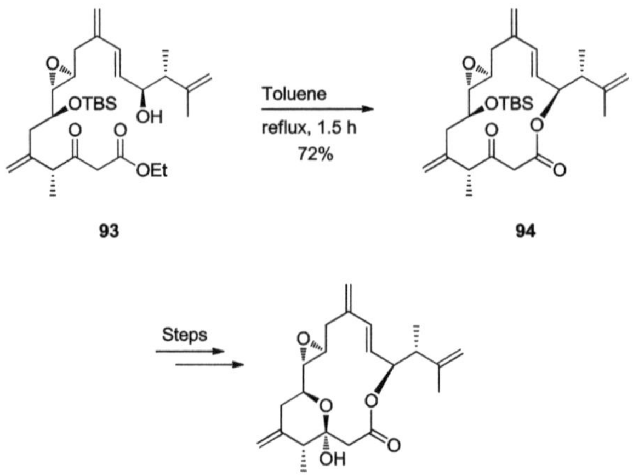

(−)-Amphidinolide P, **95**

Scheme 23 β-Keto ester thermolysis and subsequent macrolactonization in the total synthesis of (−)-amphidinolide P, **95**

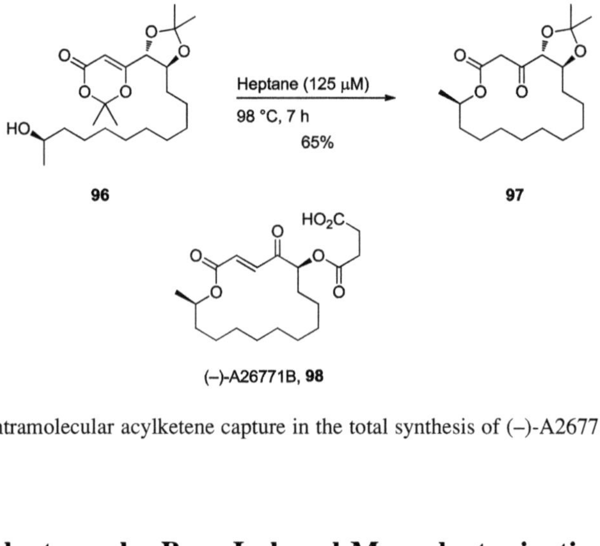

(−)-A26771B, **98**

Scheme 24 Intramolecular acylketene capture in the total synthesis of (−)-A26771B, **98**

6 Macrolactones by Base-Induced Macrolactonization of Benzodioxinones to Benzolactones

6.1 Background

A related but mechanistically different process to dioxolenone or β-keto ester thermolysis marks the base-induced macrolactonization of benzodioxinones to benzolactones (Scheme 25). This method has been independently described

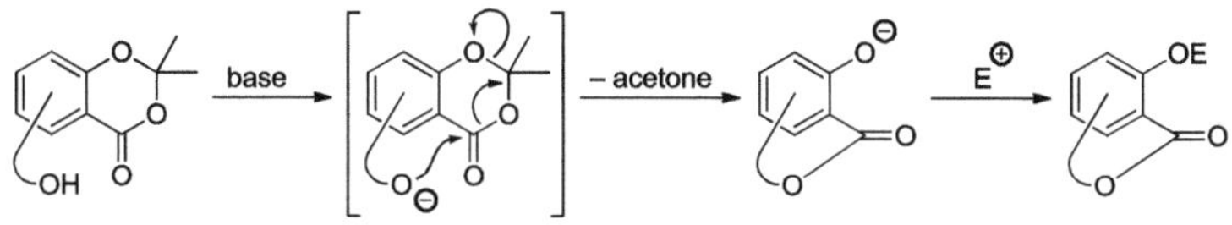

Scheme 25 Base-induced macrolactonization of benzodioxinones to benzolactones

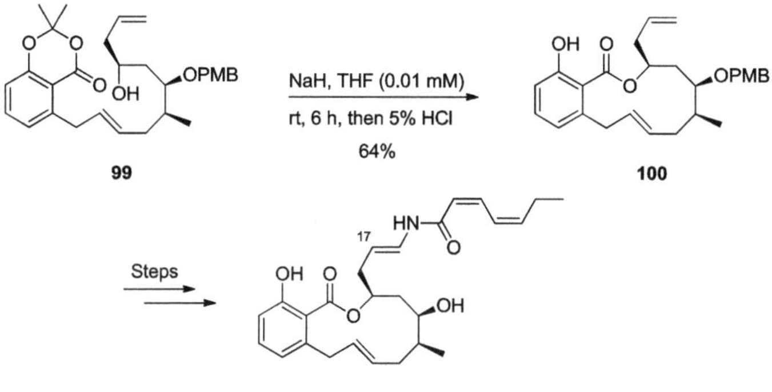

(–)-Salicylihalamide A, **65** and B, **66** (17Z-isomer)

Scheme 26 Base-induced macrolactonization during the formal total synthesis of salicylihalamide A, **65** and salicylihalamide B, **66**

by De Brabander et al. [67] and Rizzacasa et al. [68, 69] in the synthesis of benzolactone apicularens (ten-membered). It takes advantage of the loss of acetone followed by alkylation.

6.2 12-Membered Macrocyclic Lactones

6.2.1 Salicylihalamides A and B

The base-induced macrolactonization of benzodioxinones to benzolactones has been used in the synthesis of salicylihalamides A, **65** and B, **66** [70] (Scheme 26).

7 Macrolactones via Intramolecular Yamamoto et al. Vinylogous Aldol Reaction

7.1 Background

An intramolecular version of the Yamamoto vinylogous aldol reaction [71] is the macrocyclization process for the construction of 12- and 14-membered macrolides. The yields are high (up to 89 %), and the reaction can proceed with excellent remote stereocontrol (dr up to >25/1) with chiral substrates [72] (Scheme 27). This

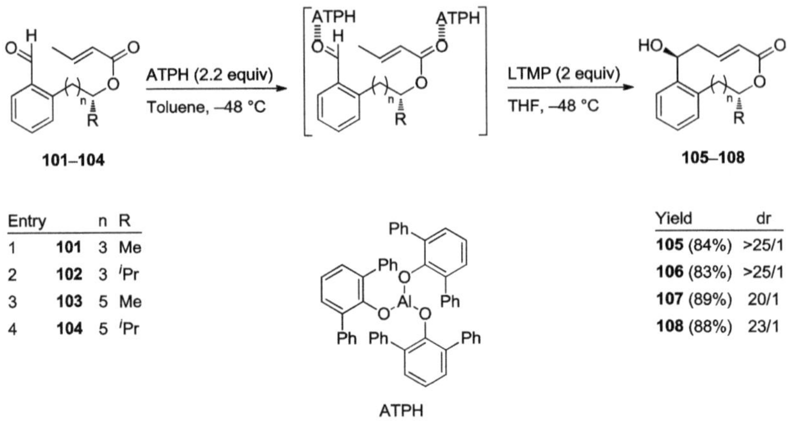

Scheme 27 Application of the intramolecular Yamamoto vinylogous aldol reaction to the synthesis of macrolides

innovative approach to direct (in terms of chemoselectivity) and discipline (in terms of regiochemistry of dienolate formation and attack) the course of the vinylogous aldol reaction is undoubtedly a prime achievement in the aldol field. The observed regioselectivity is in fact unattainable under standard conditions, and both the architecture and function of aluminium tris(2,6-diphenoxide) (ATPH) in this reaction are reminiscent of enzymes active centers.

7.2 16-Membered Macrocyclic Lactones

7.2.1 (+)-Peloruside A

The intramolecular Yamamoto vinylogous aldol reaction (lithium 2,2,6,6-tetramethyl-piperidine (LTMP) (two equivalent), aluminum tris (2,6-diphenylphenoxide) (ATPH) (2.2 equivalent), toluene/THF, –48 °C) of **109** proceeded in high yield (84 %) and good levels of diastereoselectivity (dr = 6:1) to deliver 16-membered peloruside precursor **110** [73] (Scheme 28).

8 Wittig and HWE Transformations in Macrolactone Formation

8.1 12-Membered Macrocyclic Lactones

8.1.1 (+)-Patulolide A

The Wittig and Wittig-type reactions belong to the most reliable and widely employed transformations for joining large segments during total syntheses of

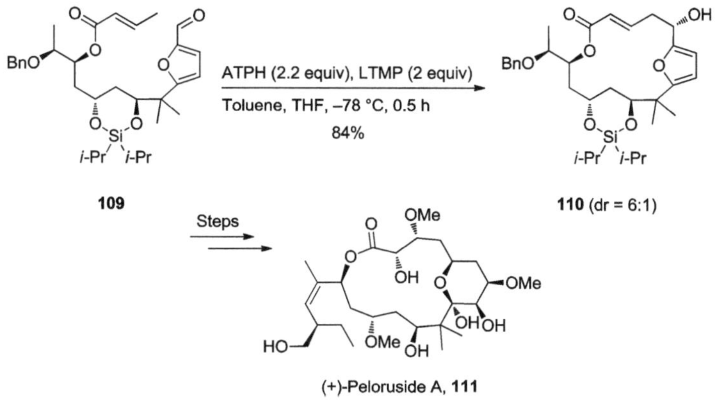

Scheme 28 Intramolecular Yamamoto vinylogous aldol reaction as a key step in the synthesis of (+)-peloruside A, **111**

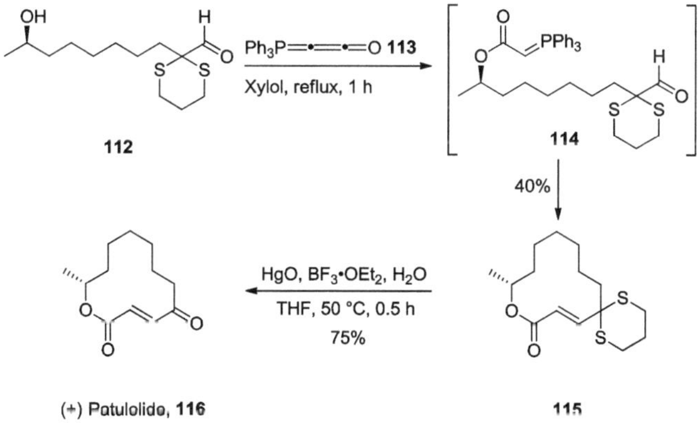

Scheme 29 Intramolecular Wittig reaction using keteneylidene triphenyl phosphorane, **113**

complex natural products. However, in recent years it became apparent that these olefinations are also suitable lactonization steps since they are compatible with a variety of functional groups and tolerate most of the employed conditions. Based on this state of knowledge, the intramolecular Wittig olefination reaction has been successfully applied in the preparation of 2*E* unsaturated 12-membered macrolides. In 1993, Bestmann et al. reported the synthesis of (+)-patulolide A **116** by this procedure [74]. The key step in this synthesis was the reaction of keteneylide triphenyl phosphorane **113** and hydroxy aldehyde **112** followed by subsequent intramolecular Wittig olefination leading to ring-closure product **115** (Scheme 29).

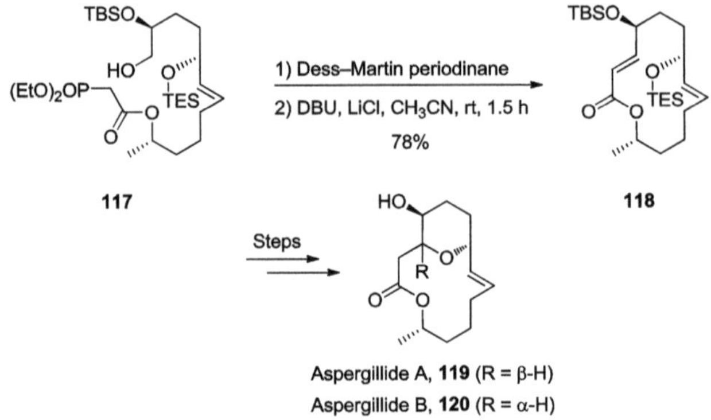

Scheme 30 Intramolecular HWE reaction in the total synthesis of aspergillide A, **119** and aspergillide B, **120**

8.2 14-Membered Macrocyclic Lactones

8.2.1 Aspergillide A and B

Oxidation of **117** with Dess–Martin periodinane and a subsequent Masamune–Roush [75] modified intramolecular Horner–Wadsworth–Emmons (HWE) reaction gave the 14-membered macrolactone **118** in 78 % yield over two steps [76] (Scheme 30).

8.3 16-Membered Macrocyclic Lactones

8.3.1 (+)-Rhizoxin D

Using the Masamune–Roush protocol [75] for base-sensitive substrates, the intramolecular Horner–Wadsworth–Emmons (HWE) coupling reaction established the macrocyclic C2–C3 bond in 80 % yield [77] (Scheme 31).

9 Stille and Stille-Type Transformations in Macrolactone Formation

Another frequently employed approach to establish ring closure is to use palladium mediated cross-coupling reactions. Among these, the Stille coupling is considered to be a promising protocol since the double-bond geometry can be established prior to the C–C bond formation and vinyl or aryl halides and stannanes are easily accessible and tolerate a variety of subsequent transformations.

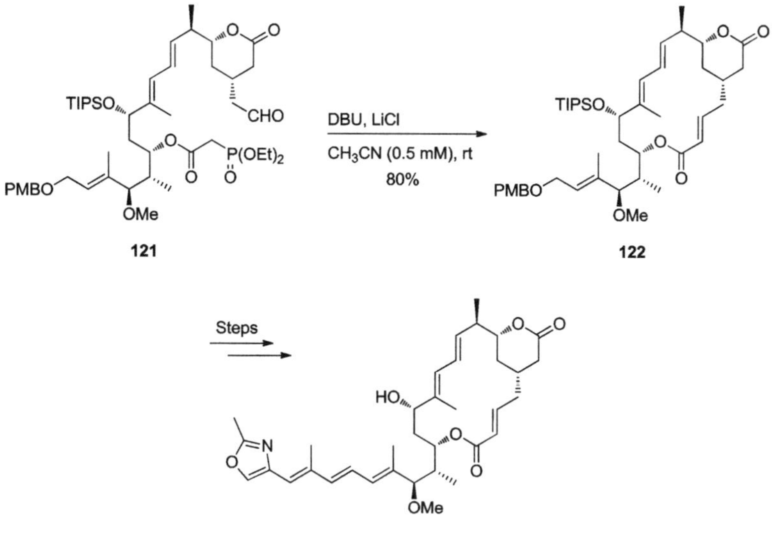

Scheme 31 Intramolecular HWE reaction in the total synthesis of (+)-rhizoxin D, **123**

9.1 14-Membered Macrocyclic Lactones

9.1.1 (*S*)-Zearalenone

The synthesis of zearalenone is an example in which the Stille coupling was employed for ring closure rather than the apparent macrolactonization. After attentive investigations, eventually the authors ascertained that by using 2–3 mol % polymer bound $Pd(PPh_3)_4$ catalyst (0.1 mmol/g polymer) in refluxing toluene, the pivotal intramolecular Stille coupling proceeded quite well in a reproducible manner (54 %) [78] (Scheme 32).

9.2 16-Membered Macrocyclic Lactones

9.2.1 Elaiolide

The study of copper-promoted Stille reaction has resulted in a number of significant new developments in the field of cross-coupling reactions. One of the most important discoveries is the use of copper(I) chloride to effect an intramolecular cyclization in the absence of any palladium catalyst [79]. This discovery has led to the rapid introduction of other copper(I) salts, such as copper(I) thiophene-2-carboxylate (CuTC) introduced by Allred and Liebeskind [80]. CuTC in *N*-methylpyrrolidinone (NMP) effects very rapid Stille cross-coupling reactions under mild

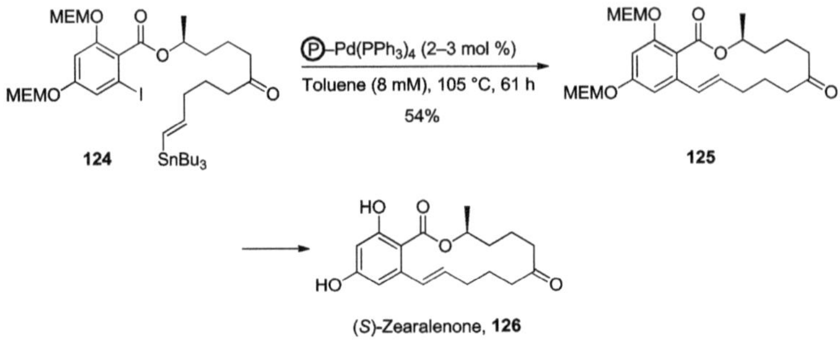

Scheme 32 Intramolecular Stille coupling using polymer bound Pd(0) catalyst $Pd(PPh_3)_4$ in the total synthesis of (*S*)-zearalenone, **126**

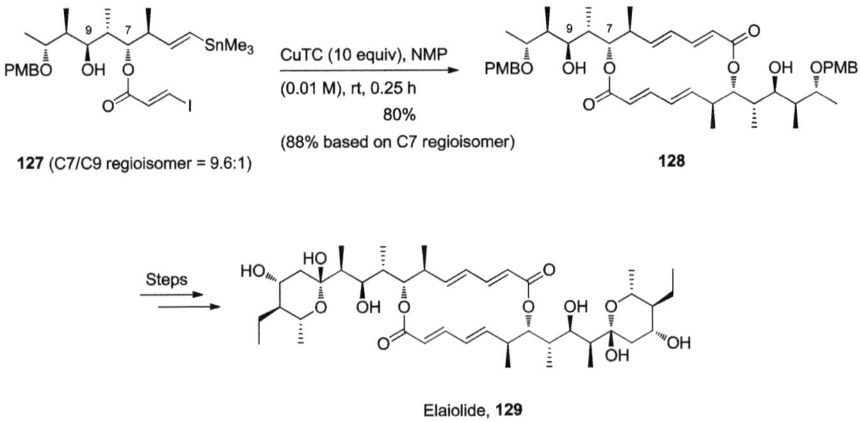

Scheme 33 Rapid intramolecular Stille cross coupling applying the Allred and Liebeskind protocol

conditions in the absence of Pd catalysis. The power of this modified Stille protocol was elegantly applied to the total synthesis of elaiolide **129** [81] (Scheme 33). Thus, treatment of a 10 mM solution of monomer **127**, in *N*-methylpyrrolidinone with CuTC (10 equivalent) at ambient temperature for 15 min, produced the required 16-membered macrocycle **128** as a white crystalline solid in excellent yield.

10 Macrolactonization Through Intramolecular Nozaki–Hiyama–Kishi Coupling

The Nozaki–Hiyama–Kishi (NHK) coupling strategy has its advantages in joining vinyl halides directly to aldehydes. The major disadvantage using this protocol is the low level of stereocontrol of the newly formed secondary alcohol. Consequently, a significant number of applications used this disconnection at positions were the so-generated alcohol has to be oxidized later on.

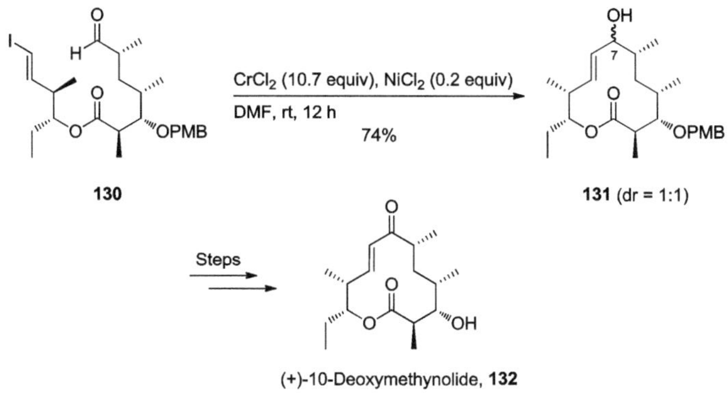

Scheme 34 Construction of 12-membered lactone **131** via intramolecular Nozaki–Hiyama–Kishi (NHK) coupling

10.1 12-Membered Macrocyclic Lactones

10.1.1 (+)-10-Deoxymethynolide

(+)-10-Deoxymethynolide **132**, the 12-membered aglycone of 10-deoxymethymycin, was efficiently constructed via an intramolecular Nozaki–Hiyama–Kishi coupling reaction. Therefore, aldehyde **130** was treated with excess of $CrCl_2$ (10.7 equivalent) containing about 2 % of $NiCl_2$ affording a 1/1 mixture of diastereomeric allylic alcohols **131**, in 74 % yield [82] (Scheme 34). Notably the stereochemistry at C7 had no consequence for the total synthesis of (+)-10-deoxymethynolide **132**, as the epimeric mixture of **131** was oxidized in the next step.

10.2 13-Membered Macrocyclic Lactones

10.2.1 (+)-Brefeldin C

The intramolecular NHK coupling reaction has also been utilized to generate 13-membered lactone **135** although with less useful levels of diastereoselection. Subjection of **133** to typical NHK conditions provided a 4/1 mixture of **134** and the desired natural product (+)-brefeldin C **135** in 60 % yield [83] (Scheme 35).

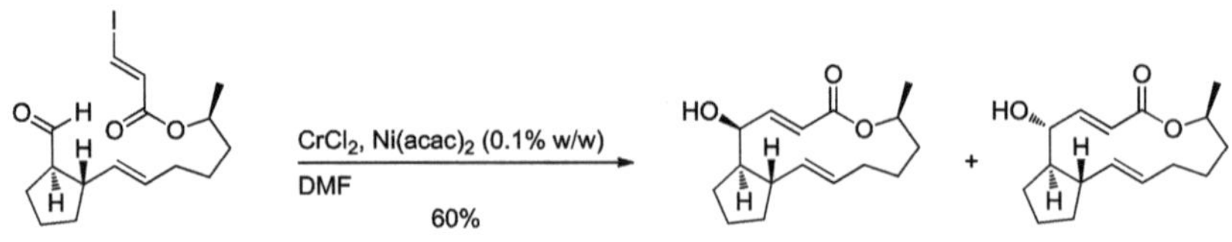

133 134 (+)-Brefeldin C, 135

(134/135 = 4:1)

Scheme 35 (+)-Brefeldin C, **135** by using an intramolecular NHK coupling

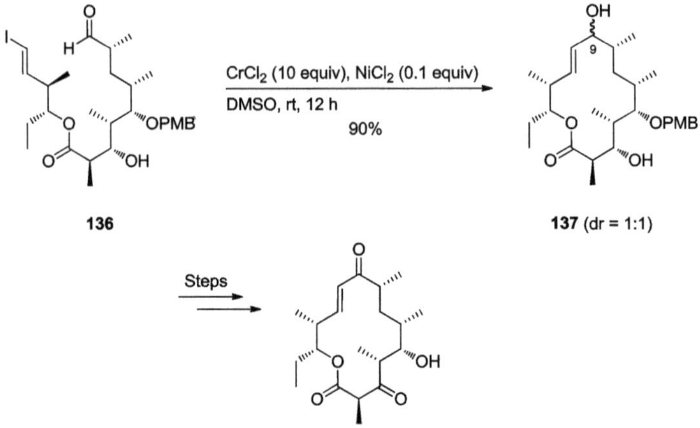

136 137 (dr = 1:1)

Narbonolide, 138

Scheme 36 Narbonolide, **138** by using an intramolecular NHK coupling

10.3 14-Membered Macrocyclic Lactones

10.3.1 Narbonolide

The intramolecular NHK coupling reaction was also used to complete the total synthesis of narbonolide **138**, a 14-membered macrolactone [84] (Scheme 36). The lack of a defined stereochemistry at C9 was nonrelevant for the total synthesis of narbonolide **138**, since compound **137** was further oxidized during the synthesis.

10.4 16-Membered Macrocyclic Lactones

10.4.1 Spiramycin I

An example for a 16-membered macrocycle using NHK coupling was presented during the aglycone synthesis of spiramycin I **141** by using a large excess of reducing $CrCl_2$ (100 equivalents) and a stoichiometric amount of $NiCl_2$. Compound

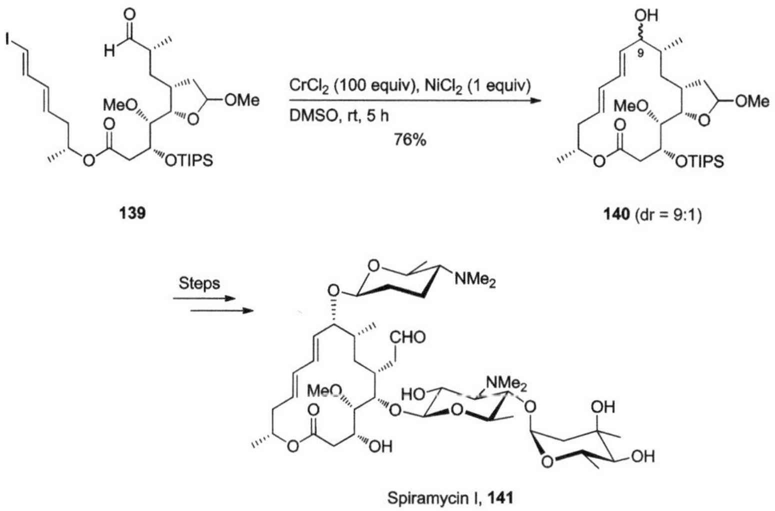

Scheme 37 Aglycone synthesis of spiramycin I, **141** by using an intramolecular NHK coupling

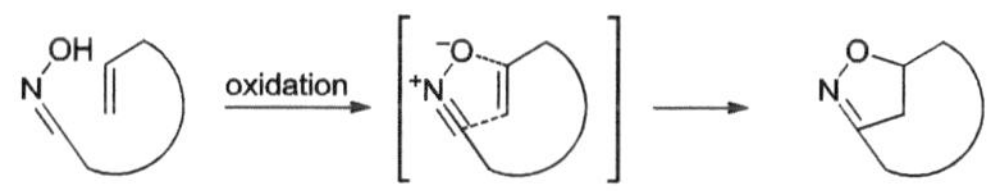

Scheme 38 Intramolecular nitrile oxide–olefin cycloaddition (INOC)

140 was formed in good yield (76 %) as a 9/1 mixture of undetermined diastereomers at C9 [85] (Scheme 37). In order to facilitate the characterization of the product, the mixture was further reacted with DDQ to give the vinylogous ketone (not shown).

11 Macrolactones via Intramolecular Nitrile Oxide–Olefin Cycloaddition

11.1 Background

Nitrile oxides, derived from oximes by oxidation with NaOCl in aqueous dioxane, undergo spontaneous 1,3-dipolar cycloadditions with olefins (e.g., acrylates) to give isoxazolines (Scheme 38). In an intramolecular setup, the intramolecular nitrile oxide–olefin cycloaddition (INOC) serves as efficient strategy for the construction of macrocyclic compounds. Remarkably, high yields, operational simplicity, and functional group tolerance of this INOC ring-closure make this strategy an attractive alternative to other ring-closing reactions.

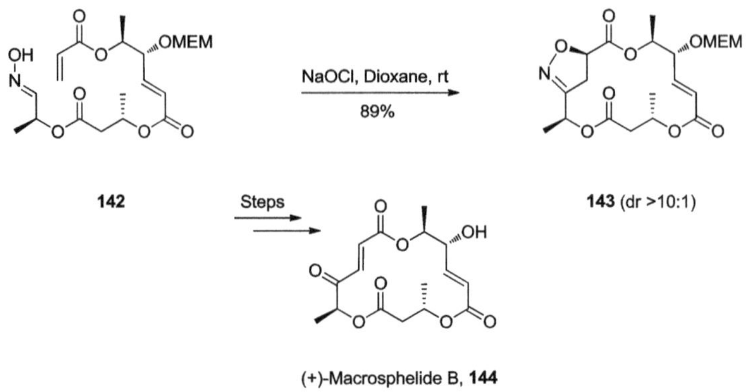

Scheme 39 INOC of aldoxime **142** in the total synthesis of (+)-macrosphelide B, **144**

11.2 16-Membered Macrocyclic Lactones

11.2.1 (+)-Macrosphelide B

In the total synthesis of (+)-macrosphelide B **144**, INOC served as an excellent pathway for the generation of new chiralities along with the ring-closure operation, since other macrocyclic ring-closure methods could not avoid undesired side reactions, such as epimerization or formation of geometric isomers [86, 87] (Scheme 39). In fact, this is an educational example of a non-aldol aldol approach and umpolung strategy since it formally joins two positively charged centers.

12 Intramolecular Late-Stage C–H Oxidative Macrolactonization

12.1 Background

The oxidative C–H macrolactonization developed by White and coworkers is proposed to proceed by an allylic C–H cleavage promoted by Pd(II)-sulfoxide [$Pd(OAc)_2 \cdot (Ph(S{=}O)CH_2)_2$ (0.3 equivalents), 1,4-benzoquinone (two equivalents) in CH_2Cl_2] to generate rapidly interconverting π-allyl-Pd(carboxylate) intermediates [88] (Scheme 40), followed by a stereodetermining C–O bond-forming event within the coordination sphere of the metal. Based on molecular modeling studies, the authors anticipated that chelate-controlled C–H macrolactonization would therefore favor formation of the desired product. Furthermore, disrupting the chelation event could provide a different stereochemical outcome by generating an earlier transition state with very little transannular character.

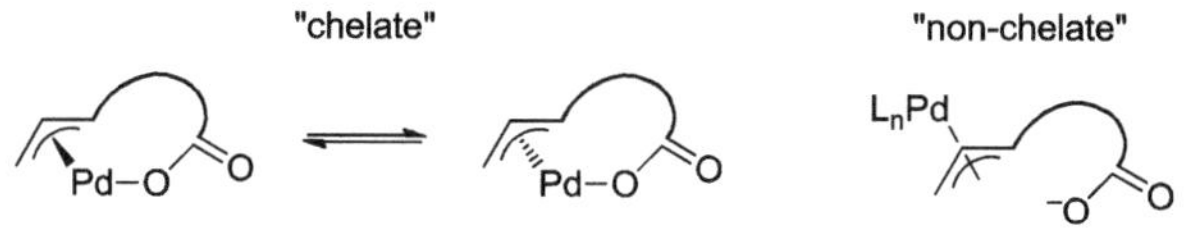

Scheme 40 Possible π-allyl-Pd(carboxylate) intermediates for C–H macrolactonization

12.2 14-Membered Macrocyclic Lactones

12.2.1 6-Deoxyerythronolide B

White and coworkers reported a late-stage C–H oxidation strategy in the total synthesis of 6-deoxyerythronolide B **147**, the aglycone precursor to erythromycin antibiotics. An advanced intermediate **145** was cyclized to give the 14-membered macrocyclic core **146** (34 % yield and 45 % recovered starting material **145**) of 6-deoxyerythronolide B **147** using a C–H oxidative macrolactonization (step 19 of 22) proceeding with high regio-, chemo-, and diastereoselectivity (dr > 40/1) [88] (Scheme 41).

13 Yamaguchi Macrolactonization

13.1 Background

The Yamaguchi protocol using 2,4,6-trichlorobenzoyl chloride **148** is probably the most popular method for performing macrolactonizations [89]. In the classical procedure (Scheme 42), the mixed anhydride is preformed in THF in the presence of triethylamine. Either with or without filtration of the $Et_3N{\cdot}HCl$ salt, since this is not pivotal in most cases [90], the mixed anhydride is diluted in toluene and slowly added by syringe pump to a highly diluted solution of DMAP (two to five equivalents) in toluene at higher temperatures (from 50 °C to reflux). The role of DMAP and related additives has been described in detail [91] and a polymer-supported DMAP reagent was employed as well [92].

13.2 12-Membered Macrocyclic Lactones

13.2.1 (+)-Lepicidin A

Since macrolactonizations are usually carried out on very advanced substrates and consequently methodological studies are rather difficult and rare, there is still no rule about the best conditions to realize a Yamaguchi macrolactonization on a

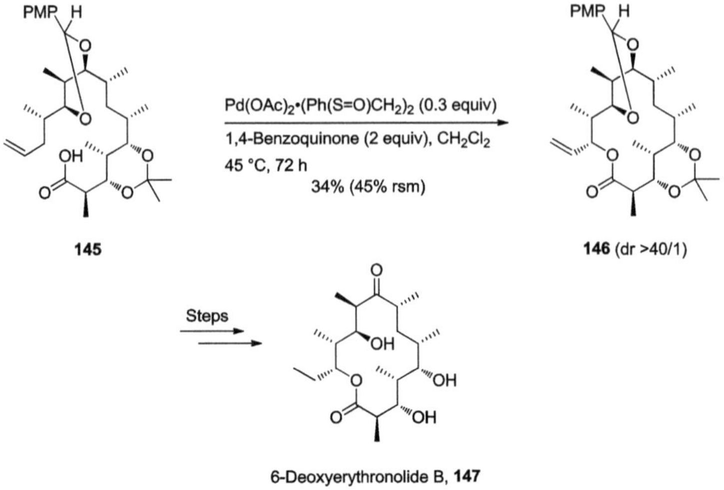

Scheme 41 Late-stage C–H oxidation strategy in the total synthesis of 6-deoxyerythronolide B, **147**

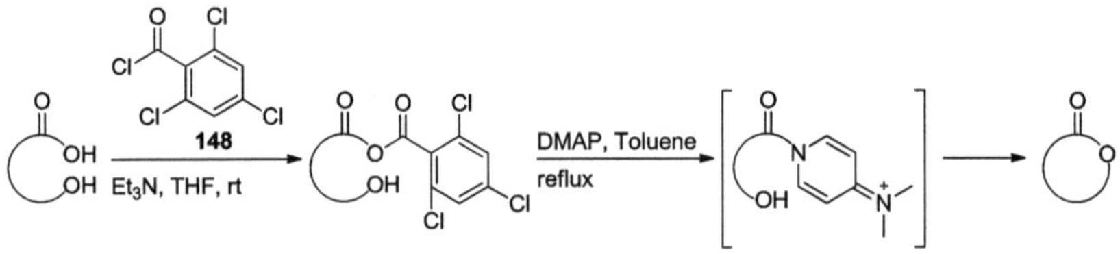

Scheme 42 Principle of the Yamaguchi macrolactonization procedure

particular substrate. For example, Evans et al. have shown the influence of the temperature and rate of addition in reducing diolide formation and destannylation in the macrolactonization of sensitive 12-membered lepicidin precursor **150** [93] (Scheme 43). Here, under optimal conditions, when rapid addition and high temperatures were combined, macrocyclization to **150** was realized in gratifying 78 % yield.

13.3 13-Membered Macrocyclic Lactones

13.3.1 Stevastelin C3

In a Yamaguchi macrolactonization approach to the stevastelins, the transformation of dihydroxy acid **152** and hydroxy acid **153** afforded both 13-membered stevastelin derivatives **154** and **155** in 82 % and 90 % yield, respectively [94] (Scheme 44). Interestingly, the formation of the corresponding 15-membered lactone was not observed.

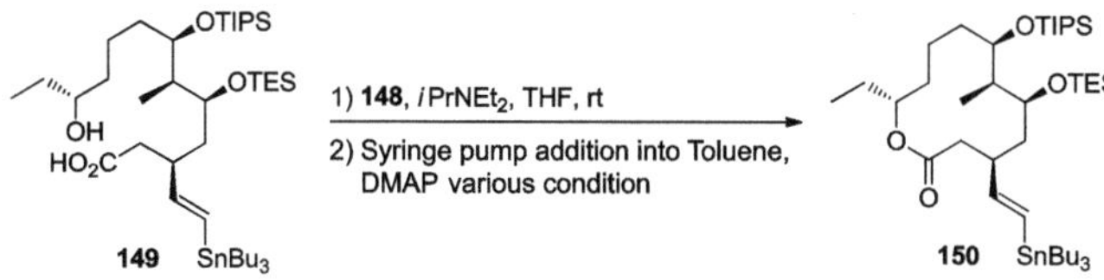

Temp (°C)	Addition Time (h)	Destannylated 149 (%)	Dimer (%)	150 (%)
25	0.5	—	33	13
70	0.5	7	10	63
110	10	63	1	10
110	1	3	3	78

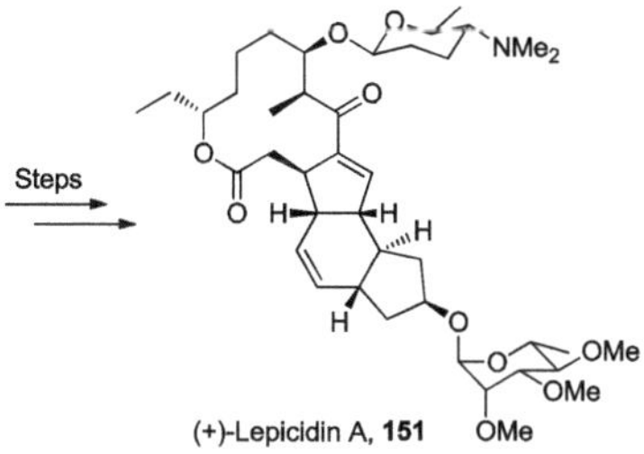

Scheme 43 Optimization of the Yamaguchi macrolactonization protocol during the synthesis of a sensitive lepicidin precursor

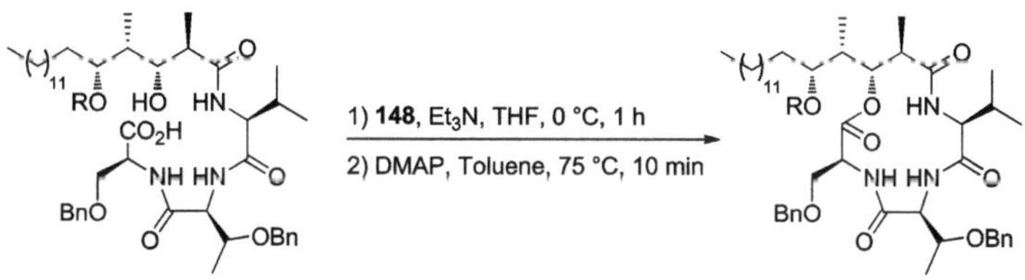

	R
152	H
153	TBS

	Yield (%)
154	82
155	90

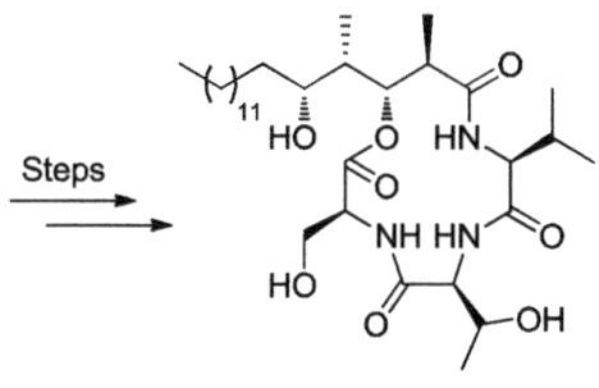

Stevastelin C3, **156**

Scheme 44 Exclusive formation of a 13-membered macrocycle using the Yamaguchi protocol

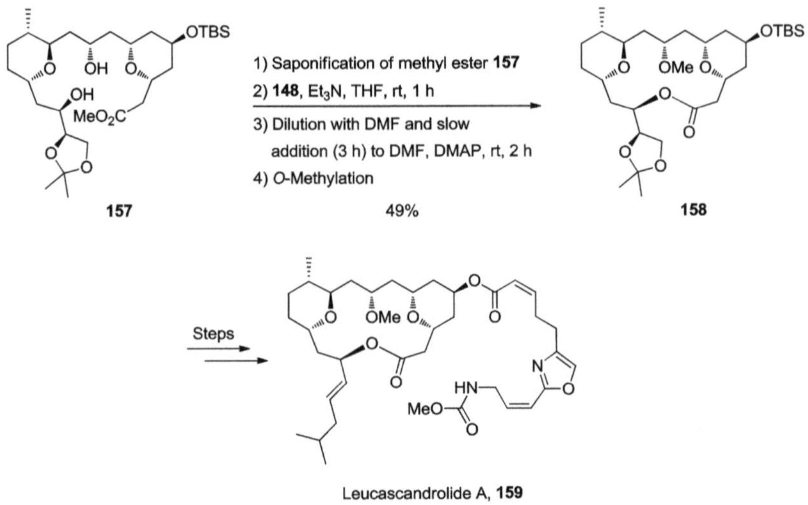

Scheme 45 Representative example of a 14-membered macrocycle using a solvent modified Yamaguchi protocol

13.4 14-Membered Macrocyclic Lactones

13.4.1 Leucascandrolide A

In the total synthesis of leucascandrolide A **159**, Carreira et al. have shown that the macrolactonization of the seco-acid of compound **157** (Scheme 45) under standard conditions led mainly to oligomeric products, probably due to unfavorable hydrogen bond interactions. To disrupt these interactions, the Yamaguchi reaction was carried out in DMF, giving the 14-membered methylated lactone **158** in 49 % overall yield and none of the undesired 8-membered lactone [95].

13.5 15-Membered Macrocyclic Lactones

13.5.1 Amphidinolide J

During the total synthesis of amphidinolide J, **164**, a 4/1 mixture of seco-acids **160** and **161** was subjected to macrolactonization under Yamaguchi conditions to afford the 15-membered macrolactone **162** (34 %) and the undesired 14-membered macrolactone **163** (24 %) which readily could be separated by flash chromatography on silica gel [96] (Scheme 46). Remarkably, **163** could be transformed to amphidinolide J **164** by intramolecular transesterification.

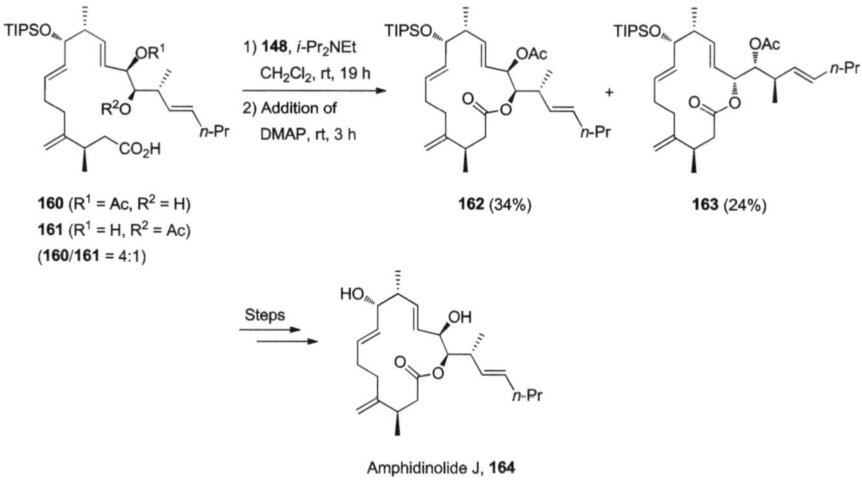

Scheme 46 Yamaguchi macrolactonization in the total synthesis of 15-membered amphidinolide J, **164**

13.6 16-Membered Macrocyclic Lactones

13.6.1 (+)-Peloruside A

The macrolactonization strategy in Jacobsen's total synthesis of (+)-peloruside A **111** [97] (Scheme 47) is similar to the former approaches to this natural product. Employing similarly protected seco-acids [98, 99], Jacobsen et al. were able to differentiate between the free hydroxy groups at C11 and C15.

14 Macrolactonization Through Thioesters: Corey (Nicolaou, Clark, Brunelle), Schmidt, Wollenberg, and Gerlach Procedures

14.1 Background

The macrolactonization of thioesters is the biosynthetic pathway for the formation of macrolides. It is therefore not surprising that this strategy was also investigated for the laboratory synthesis of macrolactones. The most prominent reaction involving a thioester is the "double activation" method described in 1974 by Corey and Nicolaou [100]. The mechanism involves the initial formation of a 2-pyridine thioester **168** of the ω-hydroxy acid by using a Mukaiyama oxidation–reduction condensation [101, 102] with PySSPy **167** and triphenylphosphine. Internal proton

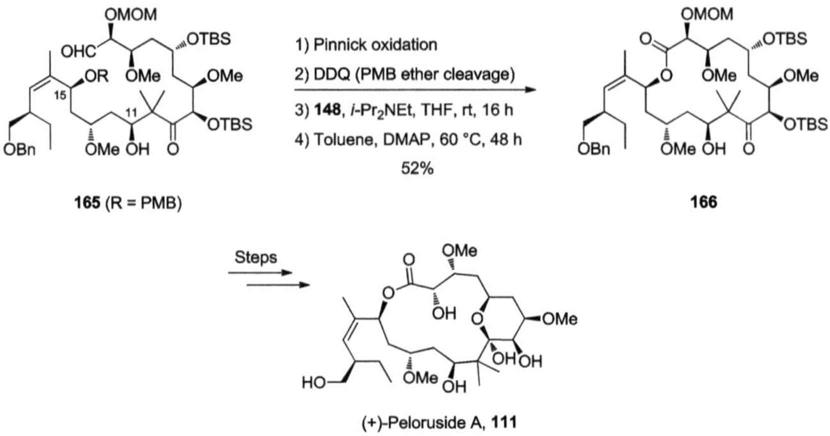

Scheme 47 Representative example of a 16-membered macrocycle using the Yamaguchi protocol

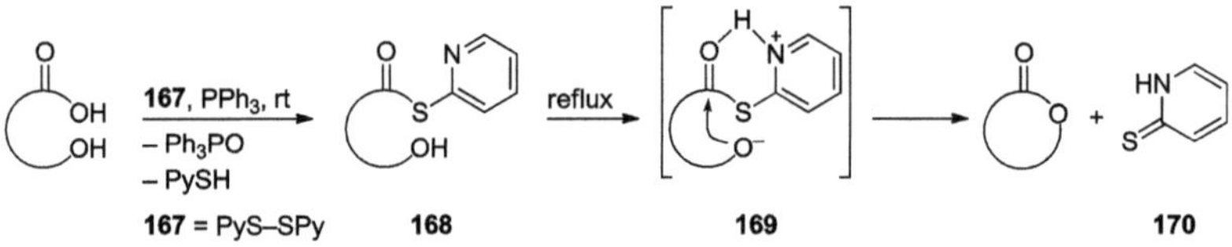

Scheme 48 The Corey–Nicolaou method

transfer then affords intermediate **169** in which both the carbonyl and the hydroxy groups have been activated, leading to the "electrostatically driven" macrolactonization product (Scheme 48).

The most frequently used reagents for thioester formation are depicted in Fig. 3 [100, 103–106]. These variants were developed to increase the yields and/or to perform the cyclization under milder conditions. For the same reasons metal ions (Ag, Cu or Hg) were used as chelating agents between the sulfur atom and the alcoholate in transition state **169**. The Gerlach and Thalmann modification introduced in 1974 is probably the most popular one [107]. It uses silver salts ($AgClO_4$, $AgBF_4$, AgOTf) and allows reactions to be carried out at ambient temperature.

14.2 12-Membered Macrocyclic Lactones

14.2.1 Enterobactin

Using the Corey–Brunelle imidazolyl disulfide **172** [104] (Fig. 3), trimer **175** could be cyclized in only 16 % yield, other methods failed or gave similar low yields. The

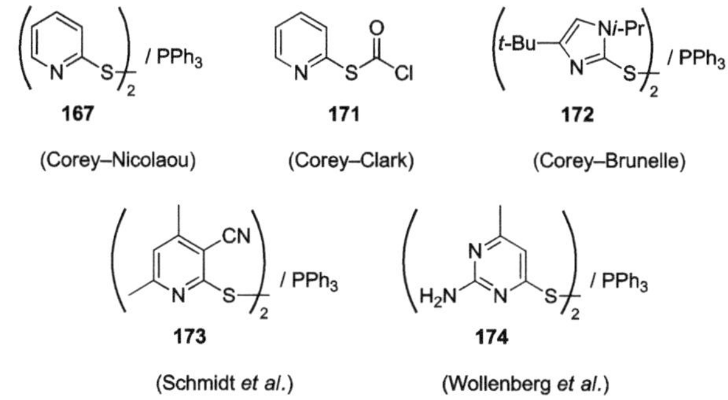

Fig. 3 Reagents for thioester formation used in macrolactonizations

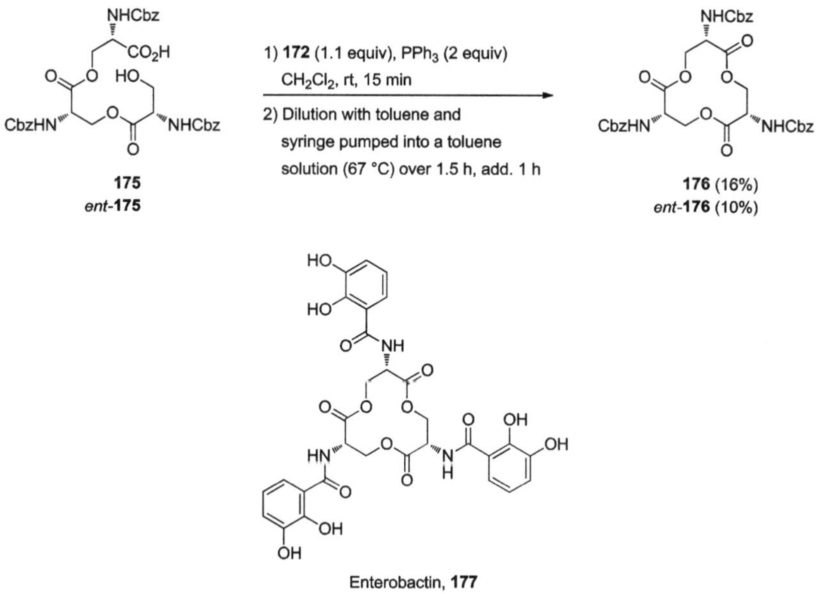

Scheme 49 Corey–Brunelle conditions in the total synthesis of enterobactin, **177**

enantiomer of **175** yielded 10 % of *ent*-**176** [108, 109] (Scheme 49). Interestingly, primarily synthesized **176** was isolated in >40 % yield also applying the Corey–Brunelle protocol indicating that this is a pivotal step [110].

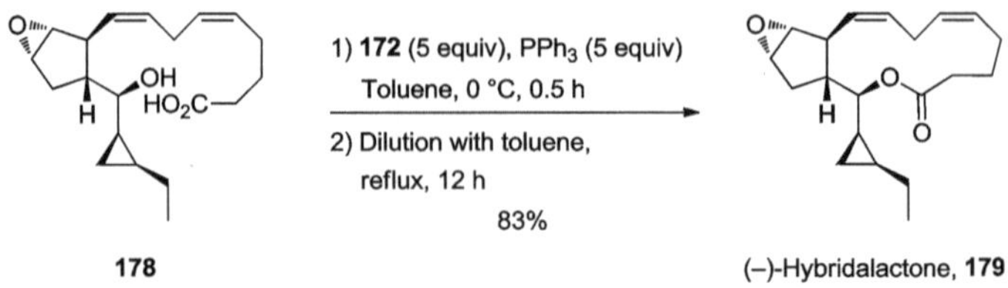

Scheme 50 Corey–Brunelle conditions in the total synthesis of (–)-hybridalactone, **179**

14.3 13-Membered Macrocyclic Lactones

14.3.1 (–)-Hybridalactone

The synthesis of (–)-hybridalactone **179**, an unusual 13-membered macrolactone with two *cis*-double bonds, was accomplished in high yield (83 %) by Corey–Brunelle lactonization [104]. Using bis(4-*tert*-butyl-*N*-isopropylimidazol-2-yl) disulfide (**172**) (Fig. 3) and triphenylphosphine (five equivalents of each) in toluene at 0 °C for 30 min, followed by dilution with toluene and heating at reflux for 12 h [111] (Scheme 50), provided the desired product, **179**.

14.4 14-Membered Macrocyclic Lactones

14.4.1 (–)-Grahamimycin A_1

One drawback of the thioester strategy can be in some cases the removal of thiopyridone **170** and triphenylphosphine oxide derived from PySSPy **167**/PPh_3 activation (Scheme 48). Based on this background, Corey and Clark [103] developed a variation of this reaction in which the thioester is synthesized first using thionylchloroformate **171** (Fig. 3). After standard workup and drying, the thioester is then directly used in the thermolysis step. This modification has been successfully applied in the total synthesis of 14-membered (–)-grahamimycin A_1, **182** [112] (Scheme 51).

14.5 16-Membered Macrocyclic Lactones

14.5.1 (+)-Macrosphelide A

A challenging macrocyclization that could only be performed using the Gerlach and Thalmann [107] protocol can be found in the total synthesis of (+)-macrosphelide A **185** [113] (Scheme 52). Here, the Corey–Nicolaou [100] protocol (PySSPy **167**/PPh_3) only provided disappointingly poor yields, and not until AgOTf was added,

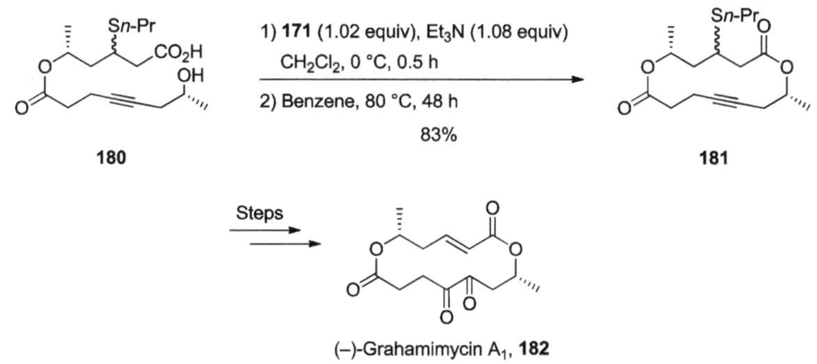

Scheme 51 Corey–Clark protocol in the total synthesis of (–)-grahamimycin A_1, **182**

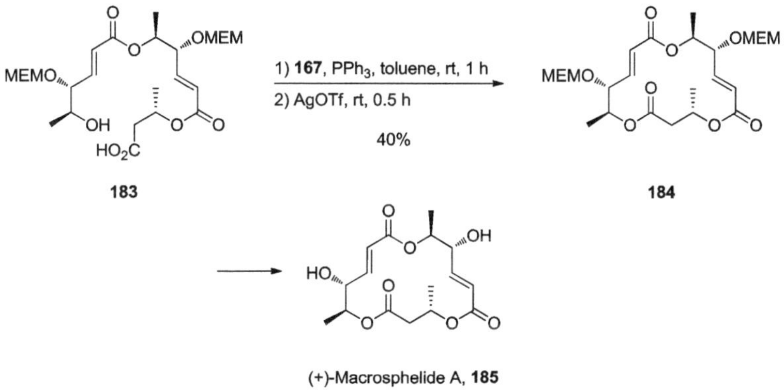

Scheme 52 The Gerlach and Thalmann protocol in the total synthesis of (+)-macrosphelide A, **185**

the macrolactone **184** could be isolated in fair 40 % yield. Noteworthy, other methods such as the Yamaguchi [89], Boden–Keck [114], or Mitsunobu [115, 116] reactions failed due to a β-elimination.

15 Mukaiyama Macrolactonization

15.1 Background

A variation of the thiopyridone concept is the Mukaiyama protocol. Here, 1-methyl-2-chloropyridinium iodide (**186**), also known as Mukaiyama's salt, is an efficient macrolactonization agent and was first described in 1976 [117] (Fig. 4). The more stable variant **187** was developed to suppress decomposition of the pyridinium salts by attack of triethylamine either on the 1-methyl group to form

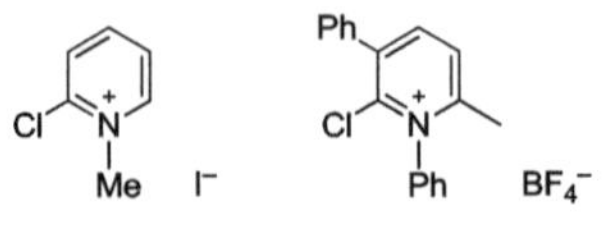

Fig. 4 Mukaiyama's reagent **186** and its further development **187**

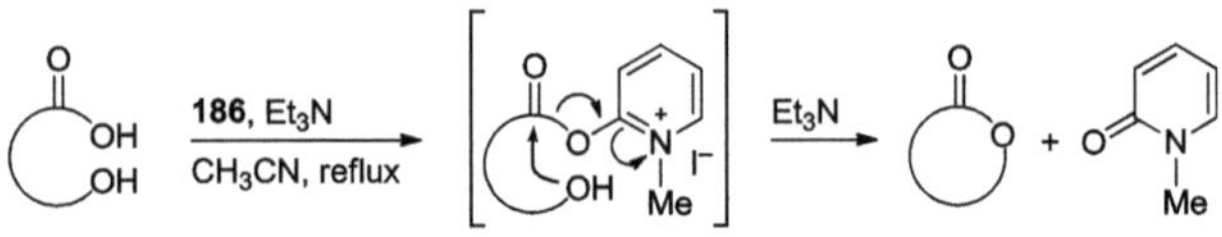

Scheme 53 Principle of the Mukaiyama macrolactonization procedure

2-chloropyridine or on the pyridinium ring to form ammoniopyridinium salts [118] (Fig. 4 and Scheme 53).

The mechanism involves chloride substitution by the carboxylate to give a highly activated acyloxypyridinium species **188** which then undergoes macrolactonization (Scheme 53).

The Mukaiyama macrolactonization is usually carried out under high-dilution conditions (10^{-3} M) with the slow addition of the hydroxy acid into a refluxing acetonitrile solution containing excess of pyridinium salt **186** and triethylamine [119].

15.2 13-Membered Macrocyclic Lactones

15.2.1 $PGF_{2\alpha}$-1,15-Lactone

The more stable pyridinium salt, 2-chloro-6-methyl-1,3-diphenylpyridinium tetrafluoroborate (**187**), was developed in order to achieve higher yields in Mukaiyama macrolactonization of complex hydroxy acids. As demonstrated in the cyclization of prostaglandin, $PGF_{2\alpha}$ **189** yielded macrocycle **190** in excellent 91 % yield [118] (Scheme 54).

15.3 14-Membered Macrocyclic Lactones

15.3.1 (–)-Gloeosporone

During the total synthesis of (–)-gloeosporone **194,** the Mukaiyama macrolactonization of diol **192** proceeded with complete chemoselectivity, 14-membered

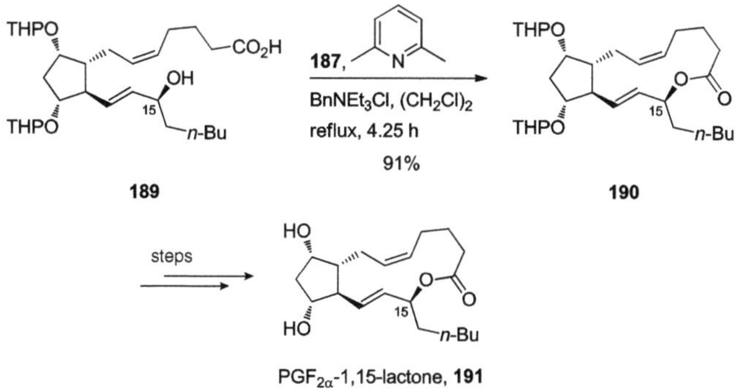

Scheme 54 An advanced Mukaiyama protocol in the total synthesis of $PGF_{2\alpha}$-1,15-lactone, **191**

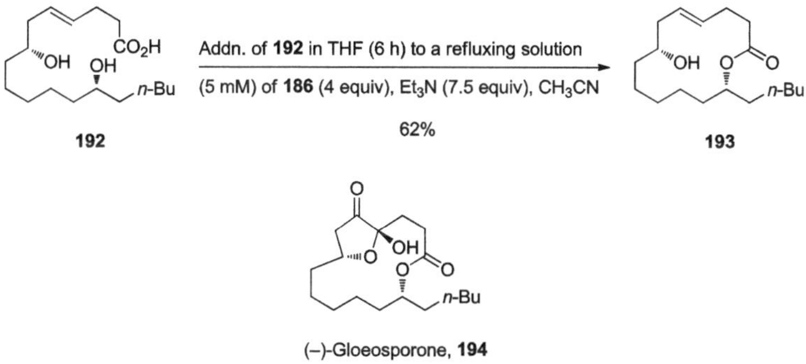

Scheme 55 The Mukaiyama protocol in the total synthesis of (–)-gloeosporone, **194**

macrolide formed rather than the eight-membered ring alternative, affording **193** in 62 % yield [120] (Scheme 55).

15.4 16-Membered Macrocyclic Lactones

15.4.1 Avermectins

Several total syntheses of avermectin B1a, **198** [121–124] and avermectin A1a, **197** [125, 126] have been reported in the literature applying the Mukaiyama macrolactonization protocol for the ring-closing key step. The most efficient one was Danishefsky's approach to avermectin A1a, **197** for which the macrolactonization was achieved in 67 % yield by using 2-chloro-*N*-methylpyridinium iodide **186** and triethylamine in methylene chloride [125, 126] (Scheme 56).

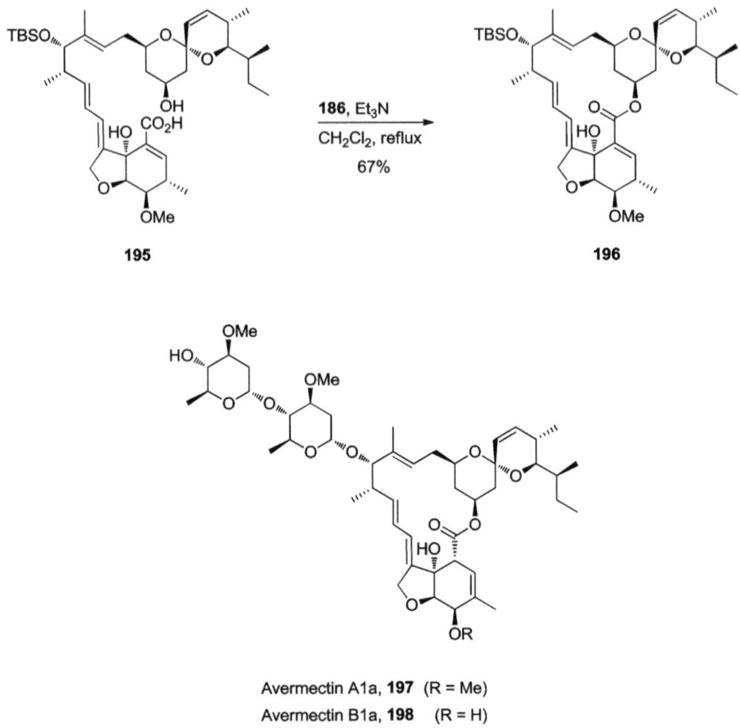

Scheme 56 The Mukaiyama protocol in the total synthesis of avermectin A1a, **197**

16 Macrolactonization Applying Phosphorus-Based Reagents

16.1 Background

Phosphates are prominent alternatives to established protocols. A variety of reagents are known such as phosphates of type **199** [127] and **200** [128], most widely used in the synthesis of esters and thiol esters [129]. The phosphorus-based peptide-coupling reagents PyBrOP [130] **201** and PyBOP [131] **202** originally designed for peptide bond formation have also been successfully used in the synthesis of macrolactones (Fig. 5).

For the first two reagents heat labile mixed carbon–phosphorus anhydrides (e.g., **203**, Scheme 57), and for the latter two species **201** and **202** acyloxyphosphonium [132] intermediates were postulated as reactive species. Since mixed carbon–phosphorus anhydrides (e.g., **203**) tend to give symmetrical anhydrides (e.g., **204**) under heating (Scheme 57), Masamune et al. recognized the need to perform macrolactonizations at a temperature below 80 °C [133].

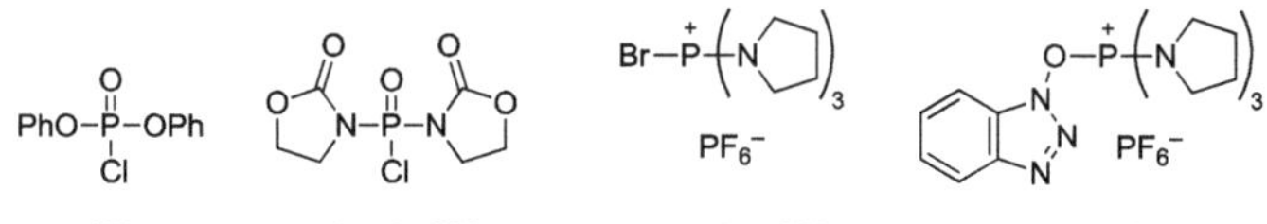

Fig. 5 Structures of frequently used phosphorus-based reagents

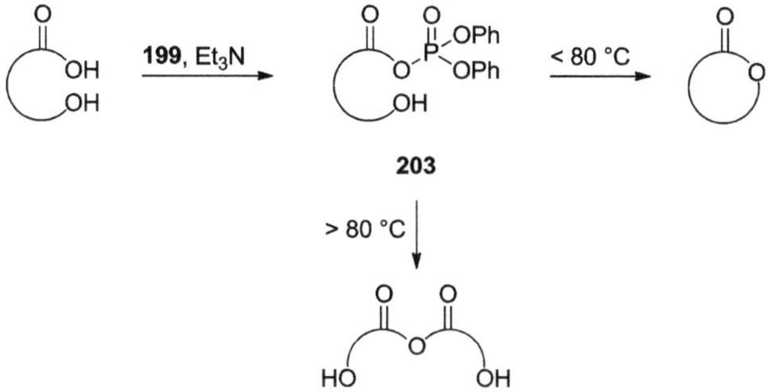

Scheme 57 Scope and limitations of phosphorus-based reagents in macrolactonization reactions

16.2 14-Membered Macrocyclic Lactones

16.2.1 (–)-Chlorothricolide

As most macrolactonizations are nontrivial, ambitious reactions and the best performance are often cumbersome which can exemplarily be seen in the total synthesis of (–)-chlorothricolide **207** [134] (Scheme 58). Not until after attentive investigations, eventually Roush and coworkers recognized that only by using Palomo–Coll's reagent BOP-Cl [128] **200** (1.9 equivalents) in toluene at 100 °C, the pivotal macrolactonization proceeded in acceptable yield (50 % of the desired macrocycle **206** and 31 % of recovered starting material **205**). On the other hand, other macrocyclization methods (Yamaguchi [89], Yamaguchi–Yonemitsu [135], Steglich [136], Boden–Keck [114], Colvin [137] and Taub–Wendler [138] methods) failed or gave unsatisfactory yields.

16.3 15-Membered Macrocyclic Lactones

16.3.1 Dynemicin A

The enediyne-bridged tricyclic core *rac*-**210** of dynemicin A **211** was obtained in 51 % yield using a PyBrOP-mediated macrolactonization followed by a

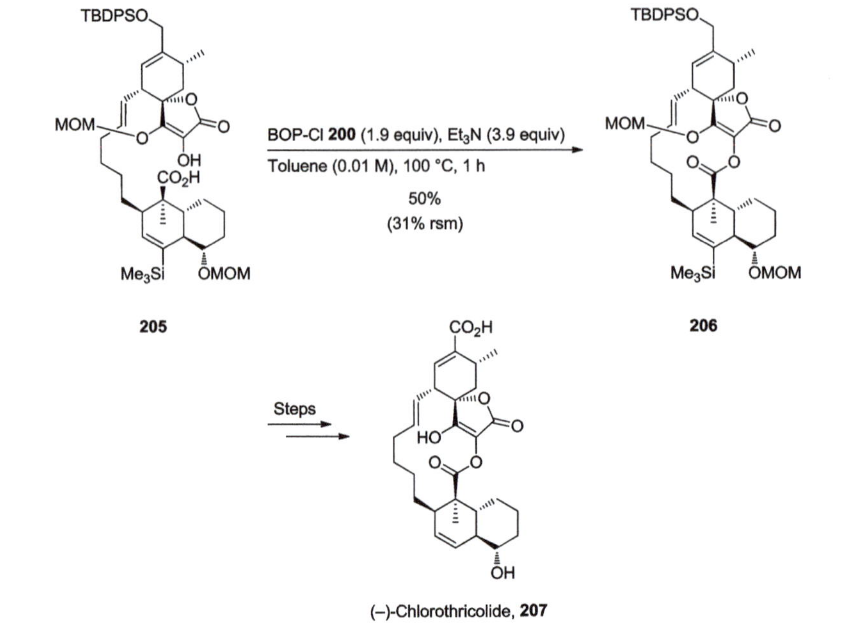

Scheme 58 BOP-Cl-mediated macrolactonization in the total synthesis of (–)-chlorothricolide, **207**

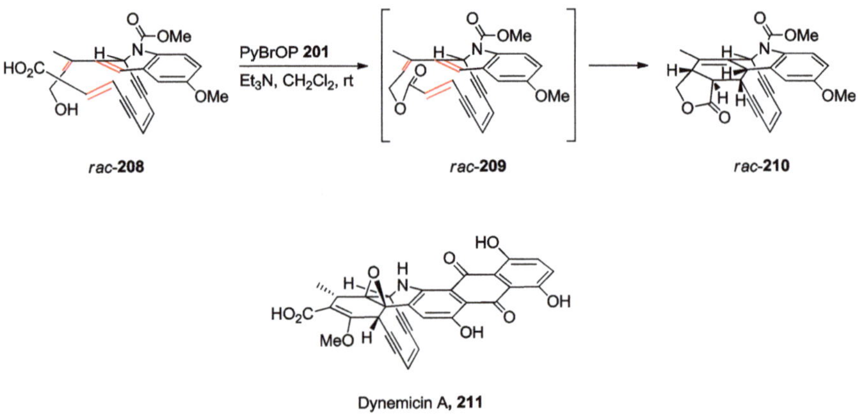

Scheme 59 PyBrOP-mediated macrolactonization followed by a transannular Diels–Alder reaction in the racemic synthesis of the enediyne-bridged tricyclic core *rac*-**210** of dynemicin A, **211**

transannular Diels–Alder reaction at room temperature [139] (Scheme 59). Although the Yamaguchi [89] macrolactonization protocol had been employed in preliminary studies [140], improved yields were only observed when the reaction was effected with PyBrOP **201** [130], which was also superior in large-scale cyclizations.

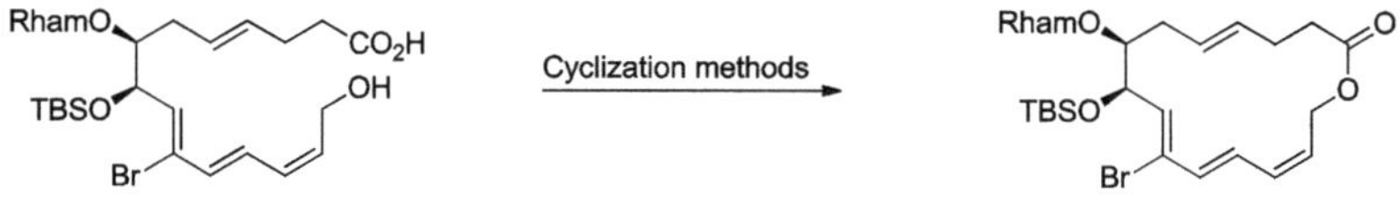

212 **213**

Entry	Method	Reagents	T (°C)	**213** (%)
1	Mukaiyama	**186**, Et_3N, CH_2Cl_2	45	33
2	Coste	**201**, DMAP, CH_2Cl_2	23	37
3	Yamaguchi–Yonemitsu	**148**, DMAP, toluene	65	38
4	Mitsunobu	DIAD, PPh_3, THF	45	48
5	Mitsunobu	DEAD, PPh_3, THF	45	54
6	Coste	**202**, DMAP, CH_2Cl_2	23	70

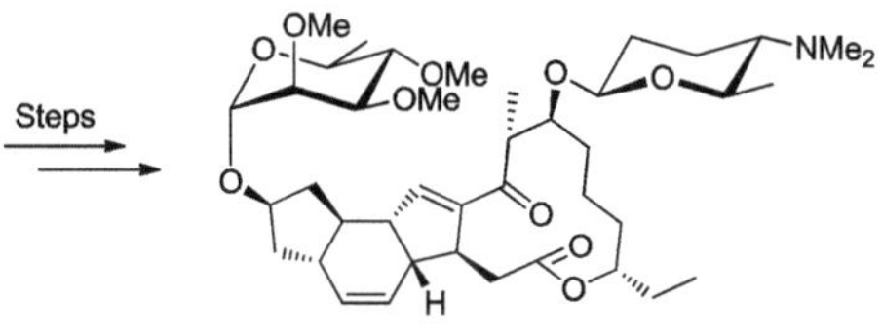

(–)-Spinosyn A, **214**

Scheme 60 Various cyclization protocols used during synthetic studies on (–)-spinosyn A, **214**

16.4 16-Membered Macrocyclic Lactones

16.4.1 (–)-Spinosyn A

Laborious investigations of the best macrolactonization strategy were imperative during synthetic studies on (–)-spinosyn A **214** (Scheme 60) [141]. Initial attempts to effect the macrocyclization of **212** by using Mukaiyama [117] conditions provided the desired lactone **213** in only 33 % yield (Scheme 60, entry 1). Coste's peptide-coupling agent PyBrOP [130] **201** and the Yonemitsu variant [135] of the Yamaguchi macrolactonization procedure provided **213** in equally low yield (37 % and 38 %) (Scheme 60, entries 2 and 3). A significant improvement (48–54 % isolated yield of **213**) was achieved when the Mitsunobu [115, 116] conditions were employed (Scheme 60, entries 4 and 5). Ultimately, the best results (70 % yield) were obtained when the macrolactonization of **212** was performed by using Coste et al. peptide-coupling agent benzotriazol-1-yloxytripyrrolidinophosphonium hexafluorophosphate (PyBOP) [131] (**202**) at ambient temperature (Scheme 60, entry 6).

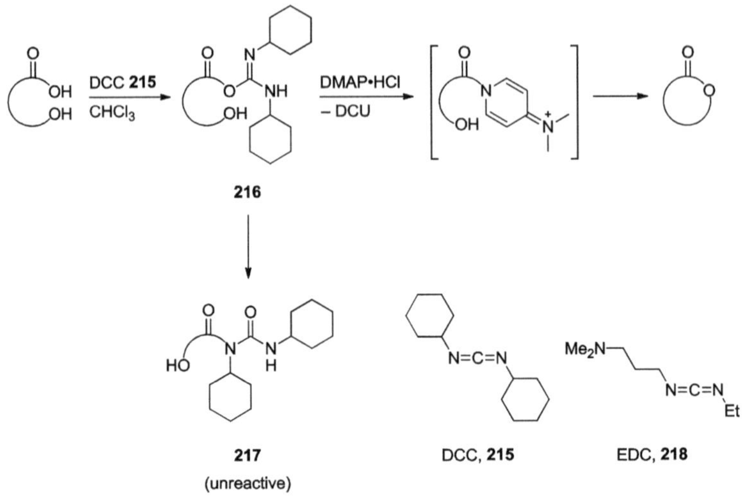

Scheme 61 The Steglich and Boden–Keck macrolactonization

17 Steglich and Boden–Keck Variants in the Synthesis of Macrocyclic Lactones

17.1 Background

Steglich's DCC–DMAP protocol has only been rarely used in macrocyclizations, mostly due to the possible formation of the unreactive *N*-acyl urea by-product **217** [136] (Scheme 61). Boden and Keck were the first who recognized the beneficial use of a catalytic amount of acid and showed this crucial role of proton transfer by using DMAP·HCl and other amine hydrochloride salts to prevent the formation of the undesired by-product **217** [114].

The main drawback associated with DCC **215**, which is usually used in excess and "quenched" by methanol in acetic acid, is the tricky removal by flash chromatography of the corresponding urea by-product, 1,3-dicyclohexylurea (DCU) (Scheme 61). Several modifications of the esterification reagent have therefore appeared, being mainly water-soluble ureas (e.g., 1-(3-dimethylaminopropyl)-3-ethylcarbodiimide (EDC) (**218**)) (Scheme 61) and supported reagents. In 2000, Keck et al. described the use of a supported DCC reagent which was easily removed by a simple filtration [142].

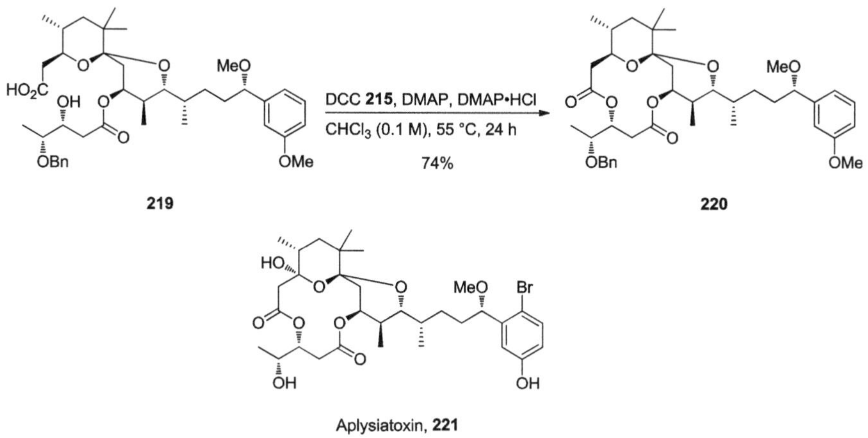

Scheme 62 The Boden–Keck protocol in the total synthesis of aplysiatoxin, **221**

17.2 12-Membered Macrocyclic Lactones

17.2.1 Aplysiatoxin

The potential of the Boden–Keck procedure was rapidly recognized and used in a large number of total syntheses, for example, in the synthesis of **220**, a 12-membered precursor of aplysiatoxin **221** [143] (Scheme 62).

17.3 13-Membered Macrocyclic Lactones

17.3.1 Diazonamide A

A remarkable example of a 13-membered macrocyclization has been documented in the synthesis of diazonamide A-related biaryl **223**. For that, phenoxy acid **222** was cyclized by treatment with Keck's modified Steglich esterification conditions to furnish lactone **223** in good yields [144] (Scheme 63).

17.4 14-Membered Macrocyclic Lactones

17.4.1 (–)-Colletol

In the first total synthesis of the unsymmetrical bis-macrolide (–)-colletol **226** [145] (Scheme 64), the penultimate step also utilized Keck's macrolactonization to assemble the 14-membered ring in excellent yields (71 % overall yield).

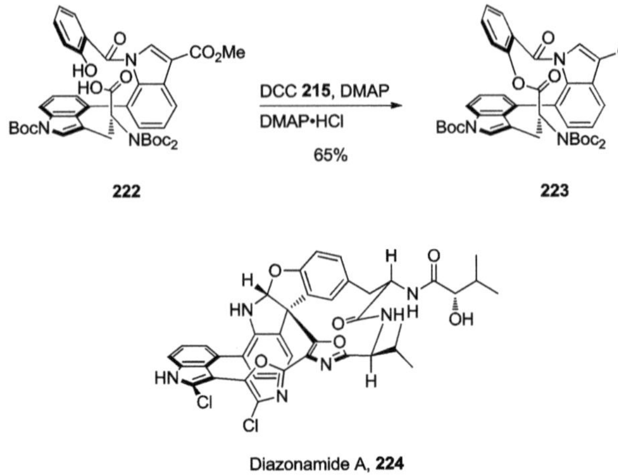

Scheme 63 The Boden–Keck protocol in the synthesis of diazonamide A-related biaryl **223**

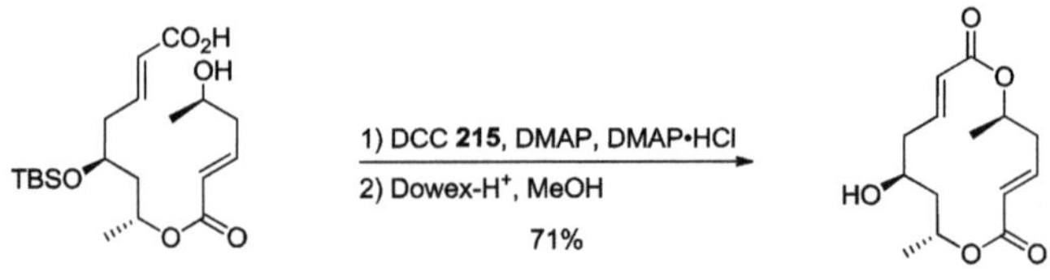

225

(–)-Colletol, 226

Scheme 64 The Boden–Keck protocol in the total synthesis of (–)-colletol, **226**

17.5 15-Membered Macrocyclic Lactones

17.5.1 (+)-Halichlorine

During the total synthesis of (+)-halichlorine **229**, the 15-membered macrocycle **228** was isolated in acceptable yield (54 %) using a EDC-modified Boden–Keck macrolactonization protocol [146] (Scheme 65).

17.6 16-Membered Macrocyclic Lactones

17.6.1 Pamamycin-607

Sometimes, DMAP can be detrimental to the macrolactonization procedure due to epimerizations and formation of intractable reaction mixtures. Therefore, in the synthesis of pamamycin-607 **232**, a Keck-type DCC/pyridine/PPTS protocol was

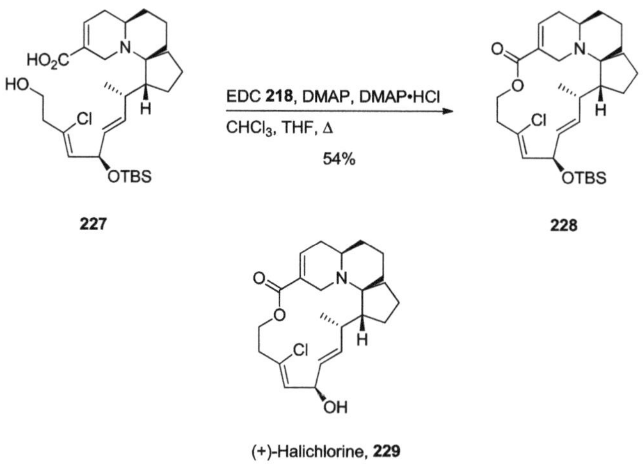

Scheme 65 EDC-modified Boden–Keck protocol in the total synthesis of (+)-halichlorine, **229**

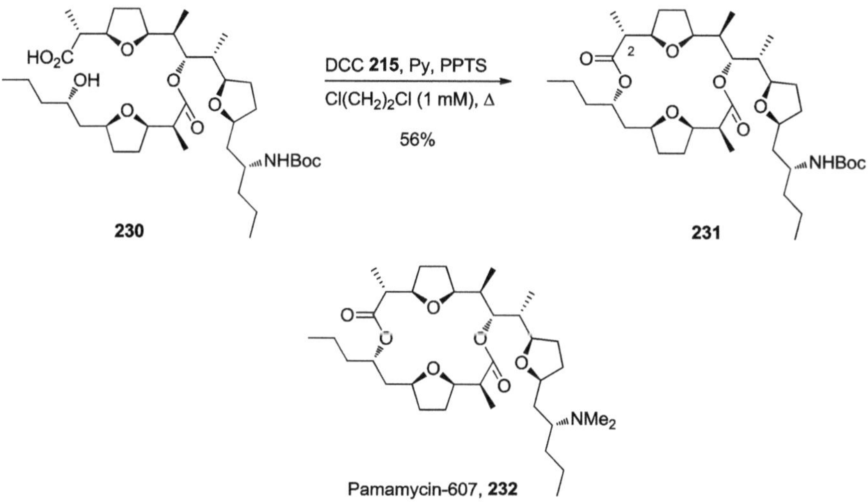

Scheme 66 The DCC/pyridine/PPTS protocol in the synthesis of pamamycin-607, **232**

the only way to give the macrodiolide in good yield [147] (Scheme 66). By contrast, Corey–Nicolaou [100], Gerlach [107], Mukaiyama [117], and Yamaguchi–Yonemitsu [135] procedures yielded no lactone, whereas the "regular" Yamaguchi [89] protocol resulted in complete epimerization of the C2 stereogenic center [147, 148].

18 The Mitsunobu Reaction in the Synthesis of Macrocyclic Lactones

18.1 Background

In 1976, Mitsunobu et al. described a macrolactonization protocol to obtain medium- and large-ring macrolactones [115, 116]. This method is based on the activation of the alcohol of the seco-acid using diethyl azodicarboxylate (DEAD) or the more hindered diisopropyl azodicarboxylate (DIAD) and triphenylphosphine. In this reaction mechanism, the key intermediate is an alkoxyphosphonium salt produced in situ, and the macrolactonization proceeds via an intramolecular S_N2 reaction and with inversion of configuration (Scheme 67). To conclude, it is worth noting that Mitsunobu reactions employing hindered alcohols can also proceed with retention of configuration.

18.2 12-Membered Macrocyclic Lactones

18.2.1 Lonomycin A

Standard Mitsunobu conditions (i.e., PPh_3/DEAD in benzene, toluene, or THF at ambient temperature) may, however, suffer from some drawbacks such as the formation of hydrazide by-products (e.g., **235**, Scheme 68). Evans et al. encountered this problem in their total synthesis of the 12-membered lonomycin A precursor, **234** (Scheme 68) and solved this issue by using the more hindered DIAD in the nonpolar solvent toluene at low temperatures [149].

18.3 13-Membered Macrocyclic Lactones

18.3.1 (+)-Brefeldin C

Another classical adoption of Mitsunobu's procedure was applied in the total synthesis of (+)-brefeldin C **135** (Scheme 69). Macrolactonization of hydroxy acid **237** yielded macrolactone **238** in excellent yield (85 %), along with a small amount of unidentified polymeric lactones (<5 %) [150].

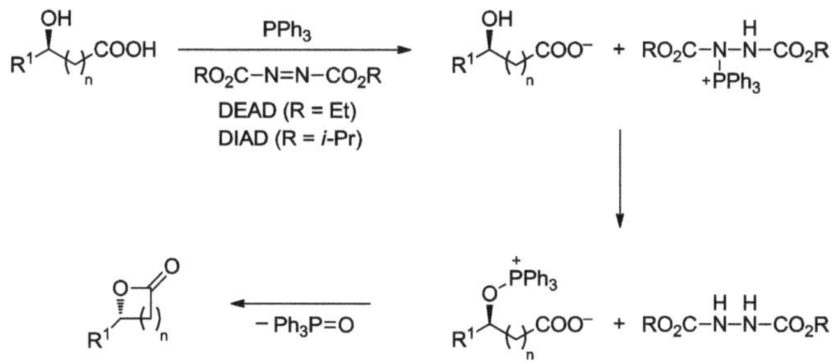

Scheme 67 The Mitsunobu macrolactonization

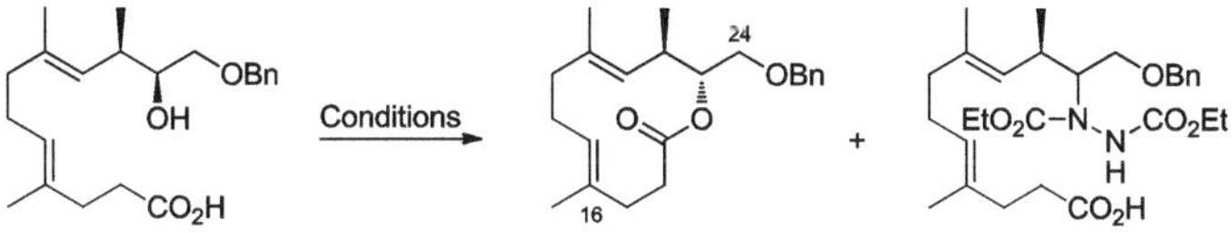

233 234 235

Reagent	Solvent	Temp (°C)	Time (h)	**234** (%)	**235** (%)
DEAD/PPh_3	THF	25	—	—	85
DEAD/PPh_3	benzene	25	—	47	—
DIAD/PPh_3	toluene	–10	0.25	95	—

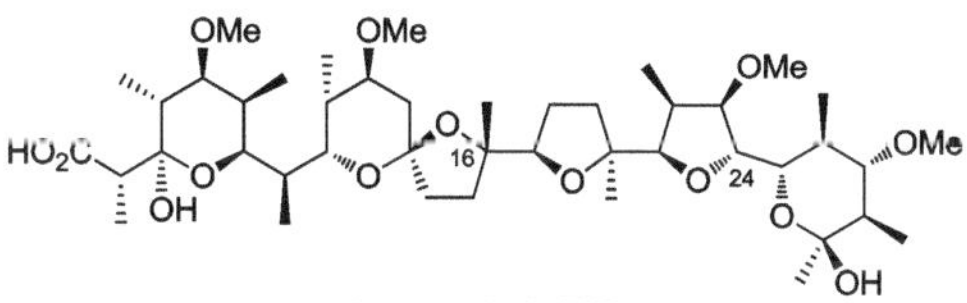

Lonomycin A, **236**

Scheme 68 The Mitsunobu protocol in the total synthesis of lonomycin A, **236**

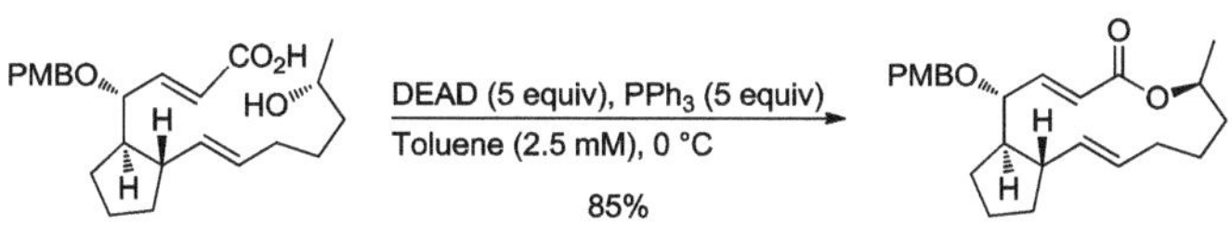

237 **238**

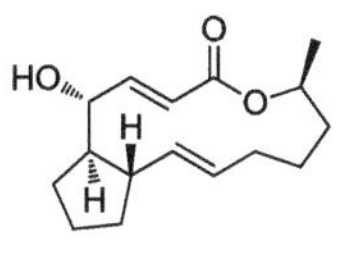

(+)-Brefeldin C, **135**

Scheme 69 The Mitsunobu protocol in the total synthesis of (+)-brefeldin C, **135**

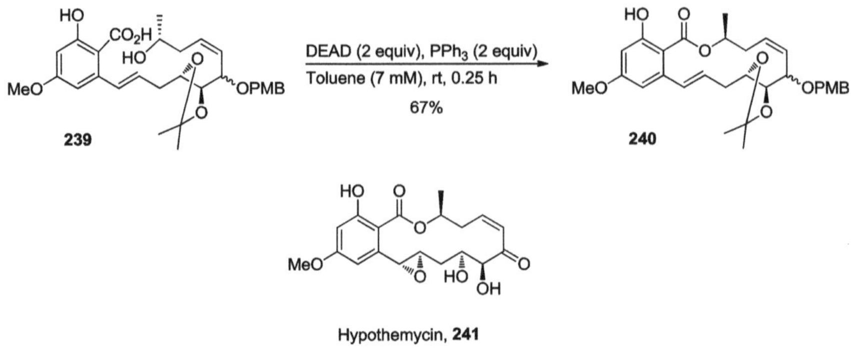

Scheme 70 The Mitsunobu protocol in the total synthesis of hypothemycin, **241**

18.4 14-Membered Macrocyclic Lactones

18.4.1 Hypothemycin

In the total synthesis of hypothemycin **241**, construction of the 14-membered ring was achieved either by using an intramolecular Suzuki coupling (15 % yield) or much more efficiently by utilizing a Mitsunobu macrolactonization, providing the identical lactone **240** in good yield (67 %) [151] (Scheme 70).

18.5 15-Membered Macrocyclic Lactones

18.5.1 Combretastatins D

The 15-membered caffrane ring of the natural product group of combretastatins D was synthesized in high yield (81 %) with suitably functionalized saturated seco-acid **242** using the Mitsunobu protocol as the key step (Scheme 71). Macrolactone **243** was then used for the construction of both combretastatins [152].

18.6 16-Membered Macrocyclic Lactones

18.6.1 FR-901,228

The use of allylic alcohols in Mitsunobu macrolactonization sometimes can be very sophisticated due to the possibility of S_N1 side reactions. This problem can be overcome by the addition of TsOH as demonstrated in the total synthesis of FR-901,228 **248** [153] (Scheme 72). Here, extensive optimization of the Mitsunobu conditions afforded lactone **247** in 62 % yield per addition of beneficial TsOH·H_2O

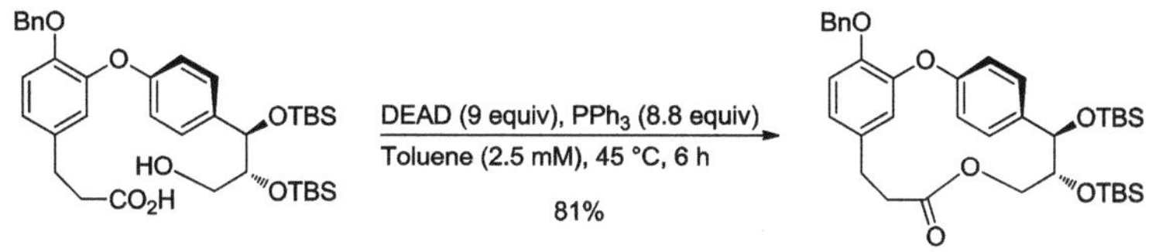

242 **243**

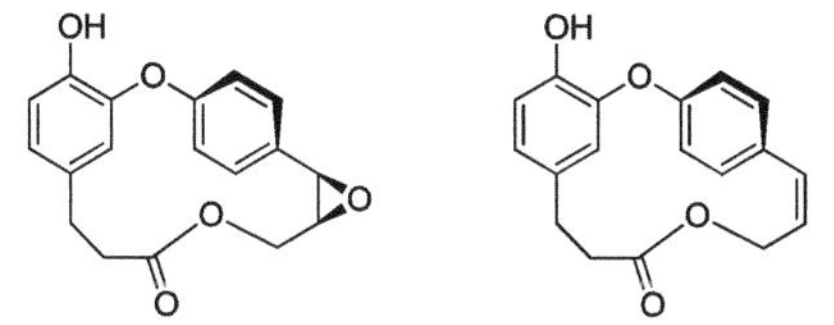

Combretastatin D-1, **244** Combretastatin D-2, **245**

Scheme 71 The Mitsunobu protocol in the total syntheses of combretastatin D-1, **244** and combretastatin D-2, **245**

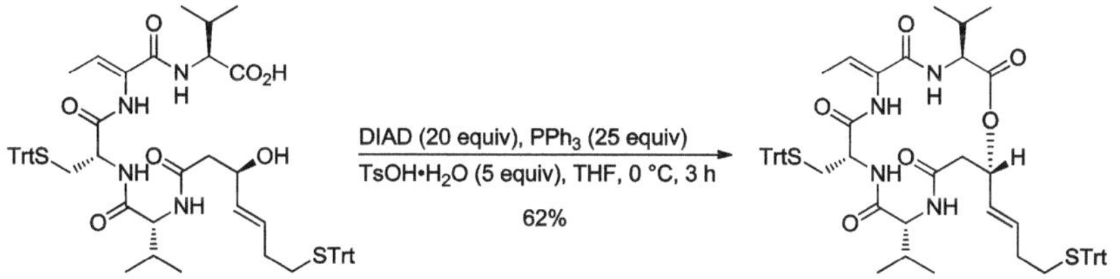

246 **247**

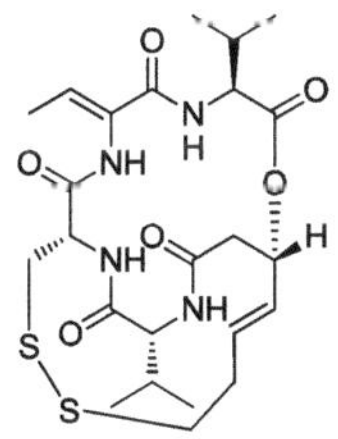

FR-901,228, **248**

Scheme 72 The Mitsunobu protocol in the total synthesis of FR-901,228, **248**

which was critical for suppressing elimination of the activated allylic alcohol. Interestingly, the (*S*)-allylic alcohol of compound **246** yielded only 5 % of the desired lactone **247** under modified Boden–Keck [114] conditions (EDC **218**, PPh_3, DMAP, DMAP·TsOH, THF).

19 Conclusion

The objective of this chapter has been to present an overview of techniques for the synthesis of 12- to 16-membered macrocyclic lactones in the total synthesis of natural products. The careful reader might have recognized that even though many other efficient macrocyclization methods such as the RCM, intramolecular cross-coupling, Nozaki–Hiyama–Kishi, and HWE methods have been developed over the years, the lactonization of seco-acids still appears to be one of the most frequently used approaches to obtain macrocyclic lactones. The need for macrolactonizations has inspired many clever solutions either by activation of the alcohol or by activation of the acid moiety and has left synthetic chemists with a variety of different strategies that allow ample of modifications in their initial retrosynthetic route.

References

1. Pennella F, Banks RL, Bailey GC (1968) Chem Commun (23): 1548
2. Mortreux A, Blanchard M (1974) J Chem Soc Chem Commun 19:786
3. Fürstner A, Stelzer F, Rumbo A, Krause H (2002) Chem Eur J 8(8):1856
4. Wengrovius JH, Sancho J, Schrock RR (1981) J Am Chem Soc 103(13):3932
5. Schrock RR, Clark DN, Sancho J, Wengrovius JH, Rocklage SM, Pedersen SF (1982) Organometallics 1(12):1645
6. Schrock RR (2002) Chem Rev 102(1):145
7. Schrock RR, Czekelius C (2007) Adv Synth Catal 349(1–2):55
8. Furstner A, Davies PW (2005) Chem Commun (18): 2307
9. Beer S, Hrib CG, Jones PG, Brandhorst K, Grunenberg J, Tamm M (2007) Angew Chem Int Ed 46(46):8890
10. Laplaza CE, Odom AL, Davis WM, Cummins CC, Protasiewicz JD (1995) J Am Chem Soc 117(17):4999
11. Laplaza CE, Cummins CC (1995) Science 268(5212):861
12. Cummins C (1998) Chem Commun (17): 1777
13. Fürstner A, Mathes C, Lehmann CW (1999) J Am Chem Soc 121(40):9453
14. Zhang W, Kraft S, Moore JS (2003) Chem Commun (7): 832
15. Zhang W, Kraft S, Moore JS (2004) J Am Chem Soc 126(1):329
16. McCullough LG, Schrock RR (1984) J Am Chem Soc 106(14):4067
17. McCullough LG, Schrock RR, Dewan JC, Murdzek JC (1985) J Am Chem Soc 107(21):5987
18. Tsai Y-C, Diaconescu PL, Cummins CC (2000) Organometallics 19(25):5260
19. Heppekausen J, Stade R, Goddard R, Fürstner A (2010) J Am Chem Soc 132(32):11045
20. Vintonyak VV, Maier ME (2007) Org Lett 9(4):655
21. Vintonyak VV, Maier ME (2007) Angew Chem Int Ed 46(27):5209
22. Vintonyak VV, Calà M, Lay F, Kunze B, Sasse F, Maier ME (2008) Chem Eur J 14(12):3709
23. Fürstner A, Bindl M, Jean L (2007) Angew Chem Int Ed 46(48):9275
24. Bindl M, Jean L, Herrmann J, Müller R, Fürstner A (2009) Chem Eur J 15(45):12310
25. Micoine K, Fürstner A (2010) J Am Chem Soc 132(40):14064
26. Fürstner A, Grela K (2000) Angew Chem Int Ed 39(7):1234
27. Fürstner A, Grela K, Mathes C, Lehmann CW (2000) J Am Chem Soc 122(48):11799
28. Fürstner A, Larionov O, Flügge S (2007) Angew Chem Int Ed 46(29):5545

29. Fürstner A, Guth O, Rumbo A, Seidel G (1999) J Am Chem Soc 121(48):11108
30. Fürstner A, Langemann K (1997) Synthesis (07): 792
31. Yang Z, He Y, Vourloumis D, Vallberg H, Nicolaou KC (1997) Angew Chem Int Ed 36 (1–2):166
32. Schinzer D, Limberg A, Bauer A, Böhm OM, Cordes M (1997) Angew Chem Int Ed 36 (5):523
33. Meng D, Bertinato P, Balog A, Su D-S, Kamenecka T, Sorensen EJ, Danishefsky SJ (1997) J Am Chem Soc 119(42):10073
34. Furstner A, Mathes C, Grela K (2001) Chem Commun (12): 1057
35. Fürstner A, Turet L (2005) Angew Chem Int Ed 44(22):3462
36. Fürstner A, De Souza D, Turet L, Fenster MDB, Parra-Rapado L, Wirtz C, Mynott R, CW L m (2007) Chem Eur J 13(1):115
37. Fürstner A, Kirk D, Fenster MDB, Aïssa C, De Souza D, Nevado C, Tuttle T, Thiel W, Müller O (2007) Chem Eur J 13(1):135
38. Deiters A, Martin SF (2004) Chem Rev 104(5):2199
39. Deshmukh PH, Blechert S (2007) Dalton Trans (24): 2479
40. Snider BB, Song F (2001) Org Lett 3(12):1817
41. Fürstner A, Dierkes T, Thiel OR, Blanda G (2001) Chem Eur J 7(24):5286
42. Smith AB III, Zheng J (2002) Tetrahedron 58(32):6455
43. Wu Y, Liao X, Wang R, Xie X-S, De Brabander JK (2002) J Am Chem Soc 124(13):3245
44. Chen J, Forsyth CJ (2004) Angew Chem Int Ed 43(16):2148
45. Ghosh AK, Xu X (2004) Org Lett 6(12):2055
46. Ferrié L, Reymond S, Capdevielle P, Cossy J (2006) Org Lett 8(16):3441
47. Zhu G, Negishi E-i (2007) Org Lett 9(15):2771
48. Chandrasekhar S, Yaragorla SR, Sreelakshmi L, Reddy CR (2008) Tetrahedron 64(22):5174
49. Gaul C, Njardarson JT, Shan D, Dorn DC, Wu K-D, Tong WP, Huang X-Y, Moore MAS, Danishefsky SJ (2004) J Am Chem Soc 126(36):11326
50. Balog A, Meng D, Kamenecka T, Bertinato P, Su D-S, Sorensen EJ, Danishefsky SJ (1996) Angew Chem Int Ed 35(23–24):2801
51. Biswas K, Lin H, Njardarson JT, Chappell MD, Chou T-C, Guan Y, Tong WP, He L, Horwitz SB, Danishefsky SJ (2002) J Am Chem Soc 124(33):9825
52. Rivkin A, Yoshimura F, Gabarda AE, Chou T-C, Dong H, Tong WP, Danishefsky SJ (2003) J Am Chem Soc 125(10):2899
53. Funk RL, Abelman MM, Jellison KM (1989) Synlett (01): 36
54. Magriotis PA, Vourloumis D, Scott ME, Tarli A (1993) Tetrahedron Lett 34(13).2071
55. Liang L, Ramaseshan M, MaGee DI (1993) Tetrahedron 49(11):2159
56. Moslin RM, Jamison TF (2006) J Am Chem Soc 128(47):15106
57. Hyatt JA, Feldman PL, Clemens RJ (1984) J Org Chem 49(26):5105
58. Boeckman RK, Pruitt JR (1989) J Am Chem Soc 111(21):8286
59. Trost BM, Gunzner JL (2001) J Am Chem Soc 123(38):9449
60. Trost BM, Gunzner JL, Dirat O, Rhee YH (2002) J Am Chem Soc 124(35):10396
61. Marshall JA, Eidam PM (2007) Org Lett 10(1):93
62. Hoye TR, Danielson ME, May AE, Zhao H (2008) Angew Chem Int Ed 47(50):9743
63. Yadav JS, Haldar A, Maity T (2012) Eur J Org Chem: 2062
64. Xue H, Yang J, Gopal P (2011) Org Lett 13(20):5696
65. Williams DR, Myers BJ, Mi L (2000) Org Lett 2(7):945
66. Gebauer J, Blechert S (2006) J Org Chem 71(5):2021
67. Bhattacharjee A, De Brabander JK (2000) Tetrahedron Lett 41(42):8069
68. Hilli F, White JM, Rizzacasa MA (2002) Tetrahedron Lett 43(47):8507
69. Hilli F, White JM, Rizzacasa MA (2004) Org Lett 6(8):1289
70. Holloway GA, Hügel HM, Rizzacasa MA (2003) J Org Chem 68(6):2200
71. Saito S, Shiozawa M, Ito M, Yamamoto H (1998) J Am Chem Soc 120(4):813
72. Abramite JA, Sammakia T (2007) Org Lett 9(11):2103

73. Gazaille JA, Abramite JA, Sammakia T (2012) Org Lett 14(1):178
74. Bestmann HJ, Kellermann W, Pecher B (1993) Synthesis (01): 149
75. Blanchette MA, Choy W, Davis JT, Essenfeld AP, Masamune S, Roush WR, Sakai T (1984) Tetrahedron Lett 25(21):2183
76. Kanematsu M, Yoshida M, Shishido K (2011) Angew Chem Int Ed 50(11):2618
77. Jiang Y, Hong J, Burke SD (2004) Org Lett 6(9):1445
78. Kalivretenos A, Stille JK, Hegedus LS (1991) J Org Chem 56(8):2883
79. Piers E, Wong T (1993) J Org Chem 58(14):3609
80. Allred GD, Liebeskind LS (1996) J Am Chem Soc 118(11):2748
81. Paterson I, Lombart H-G, Allerton C (1999) Org Lett 1(1):19
82. Pilli RA, de Andrade CKZ, Souto CRO, de Meijere A (1998) J Org Chem 63(22):7811
83. Schreiber SL, Meyers HV (1988) J Am Chem Soc 110(15):5198
84. Venkatraman L, Salomon CE, Sherman DH, Fecik RA (2006) J Org Chem 71(26):9853
85. Oddon G, Uguen D (1998) Tetrahedron Lett 39(10):1157
86. Paek S-M, Seo S-Y, Kim S-H, Jung J-W, Lee Y-S, Jung J-K, Suh Y-G (2005) Org Lett 7 (15):3159
87. Paek S-M, Suh Y-G (2011) Molecules 16(6):4850
88. Stang EM, Christina White M (2009) Nat Chem 1(7):547
89. Inanaga J, Hirata K, Saeki H, Katsuki T, Yamaguchi M (1979) Bull Chem Soc Jpn 52(7):1989
90. Evans DA, Connell BT (2003) J Am Chem Soc 125(36):10899
91. Spivey AC, Arseniyadis S (2004) Angew Chem Int Ed 43(41):5436
92. Storer RI, Takemoto T, Jackson PS, Brown DS, Baxendale IR, Ley SV (2004) Chem Eur J 10 (10):2529
93. Evans DA, Black WC (1993) J Am Chem Soc 115(11):4497
94. Sarabia F, Chammaa S, López-Herrera FJ (2002) Tetrahedron Lett 43(16):2961
95. Fettes A, Carreira EM (2003) J Org Chem 68(24):9274
96. Barbazanges M, Meyer C, Cossy J (2008) J Org Lett 10(20):4489
97. McGowan MA, Stevenson CP, Schiffler MA, Jacobsen EN (2010) Angew Chem Int Ed 49 (35):6147
98. Evans DA, Welch DS, Speed AWH, Moniz GA, Reichelt A, Ho S (2009) J Am Chem Soc 131 (11):3840
99. Ghosh AK, Xu X, Kim J-H, Xu C-X (2008) Org Lett 10(5):1001
100. Corey EJ, Nicolaou KC (1974) J Am Chem Soc 96(17):5614
101. Endo T, Ikenaga S, Mukaiyama T (1970) Bull Chem Soc Jpn 43(8):2632
102. Mukaiyama T (1976) Angew Chem Int Ed Engl 15(2):94
103. Corey EJ, Clark DA (1979) Tetrahedron Lett 20(31):2875
104. Corey EJ, Brunelle DJ (1976) Tetrahedron Lett 17(38):3409
105. Schmidt U, Heermann D (1979) Angew Chem Int Ed Engl 18(4):308
106. Nimitz JS, Wollenberg RH (1978) Tetrahedron Lett 19(38):3523
107. Gerlach H, Thalmann A (1974) Helv Chim Acta 57(8):2661
108. Rastetter WH, Erickson TJ, Venuti MC (1980) J Org Chem 45(24):5011
109. Rastetter WH, Erickson TJ, Venuti MC (1981) J Org Chem 46(18):3579
110. Corey EJ, Bhattacharyya S (1977) Tetrahedron Lett 18(45):3919
111. Corey EJ, De B (1984) J Am Chem Soc 106(9):2735
112. Hillis LR, Ronald RC (1985) J Org Chem 50(4):470
113. Kusaka S-i, Dohi S, Doi T, Takahashi T (2003) Tetrahedron Lett 44(49):8857
114. Boden EP, Keck GE (1985) J Org Chem 50(13):2394
115. Kurihara T, Nakajima Y, Mitsunobu O (1976) Tetrahedron Lett 17(28):2455
116. Mitsunobu O (1981) Synthesis (01): 1
117. Mukaiyama T, Usui M, Saigo K (1976) Chem Lett 5(1):49
118. Narasaka K, Maruyama K, Mukaiyama T (1978) Chem Lett 7(8):885
119. Mukaiyama T (1979) Angew Chem Int Ed Engl 18(10):707

120. Schreiber SL, Kelly SE, Porco JA, Sammakia T, Suh EM (1988) J Am Chem Soc 110 (18):6210
121. Armstrong A, Ley SV, Madin A, Mukherjee S (1990) Synlett 1990(06):328
122. Ley SV, Armstrong A, Diez-Martin D, Ford MJ, Grice P, Knight JG, Kolb HC, Madin A, Marby CA, Mukherjee S, Shaw AN, Slawin AMZ, Vile S, White AD, Williams DJ, Woods M (1991) J Chem Soc Perkin Trans 1(4):667
123. White JD, Bolton GL (1990) J Am Chem Soc 112(4):1626
124. White JD, Bolton GL, Dantanarayana AP, Fox CMJ, Hiner RN, Jackson RW, Sakuma K, Warrier US (1995) J Am Chem Soc 117(7):1908
125. Danishefsky SJ, Armistead DM, Wincott FE, Selnick HG, Hungate R (1987) J Am Chem Soc 109(26):8117
126. Danishefsky SJ, Armistead DM, Wincott FE, Selnick HG, Hungate R (1989) J Am Chem Soc 111(8):2967
127. Masamune S, Kamata S, Diakur J, Sugihara Y, Bates GS (1975) Can J Chem 53(23):3693
128. Diago-Meseguer J, Palomo-Coll AL, Fernández-Lizarbe JR, Zugaza Bilbao A (1980) Syn thesis (07): 547
129. Arrieta A, García T, Lago JM, Palomo C (1983) Synth Commun 13(6):471
130. Coste J, Frérot E, Jouin P, Castro B (1991) Tetrahedron Lett 32(17):1967
131. Frérot E, Coste J, Pantaloni A, Dufour M-N, Jouin P (1991) Tetrahedron 47(2):259
132. Coste J, Frerot E, Jouin P (1994) J Org Chem 59(9):2437
133. Kaiho T, Masamune S, Toyoda T (1982) J Org Chem 47(8):1612
134. Roush WR, Sciotti RJ (1998) J Am Chem Soc 120(30):7411
135. Hikota M, Tone H, Horita K, Yonemitsu O (1990) J Org Chem 55(1):7
136. Neises B, Steglich W (1978) Angew Chem Int Ed Engl 17(7):522
137. Colvin EW, Purcell TA, Raphael RA (1976) J Chem Soc Perkin Trans 1(16):1718
138. Taub D, Girotra NN, Hoffsommer RD, Kuo CH, Slates HL, Weber S, Wendler NL (1968) Tetrahedron 24(6):2443
139. Wood JL, Porco JA, Taunton J, Lee AY, Clardy J, Schreiber SL (1992) J Am Chem Soc 114 (14):5898
140. Porco JA, Schoenen FJ, Stout TJ, Clardy J, Schreiber SL (1990) J Am Chem Soc 112 (20):7410
141. Frank SA, Roush WR (2002) J Org Chem 67(12):4316
142. Keck GE, Sanchez C, Wager CA (2000) Tetrahedron Lett 41(45):8673
143. Ireland RE, Thaisrivongs S, Dussault PH (1988) J Am Chem Soc 110(17):5768
144. Feldman KS, Eastman KJ, Lessene G (2002) Org Lett 4(20):3525
145. Keck GE, Murry JA (1991) J Org Chem 56(23):6606
146. Trauner D, Schwarz JB, Danishefsky SJ (1999) Angew Chem Int Ed 38(23):3542
147. Jeong EJ, Kang EJ, Sung LT, Hong SK, Lee E (2002) J Am Chem Soc 124(49):14655
148. Lee E, Jeong EJ, Kang EJ, Sung LT, Hong SK (2001) J Am Chem Soc 123(41):10131
149. Evans DA, Ratz AM, Huff BE, Sheppard GS (1995) J Am Chem Soc 117(12):3448
150. Hatakeyama S, Osanai K, Numata H, Takano S (1989) Tetrahedron Lett 30(36):4845
151. Sellès P, Lett R (2002) Tetrahedron Lett 43(26):4627
152. Couladouros EA, Soufli IC, Moutsos VI, Chadha RK (1998) Chem Eur J 4(1):33
153. Li KW, Wu J, Xing W, Simon JA (1996) J Am Chem Soc 118(30):7237

Index

J. Cossy (ed.), *Synthesis of Saturated Oxygenated Heterocycles II*, Topics in Heterocyclic Chemistry 36, DOI 10.1007/978-3-642-41470-1,

M

N

O

MIX
Papier aus verantwortungsvollen Quellen
Paper from responsible sources
FSC® C105338

If you have any concerns about our products,
you can contact us on
ProductSafety@springernature.com

In case Publisher is established outside the EU,
the EU authorized representative is:
Springer Nature Customer Service Center GmbH
Europaplatz 3, 69115 Heidelberg, Germany

Printed by Libri Plureos GmbH
in Hamburg, Germany